AF457797

aus der Reihe:

Innovationen mit Mikrowellen und Licht

Forschungsberichte aus dem Ferdinand-Braun-Institut, Leibniz-Institut für Höchstfrequenztechnik

Band 59

Anissa Zeghuzi

Analysis of Spatio-Temporal Phenomena in High-Brightness Diode Lasers using Numerical Simulations

Herausgeber: Prof. Dr. Günther Tränkle, Prof. Dr.-Ing. Wolfgang Heinrich

Ferdinand-Braun-Institut
Leibniz-Institut
für Höchstfrequenztechnik (FBH)
Gustav-Kirchhoff-Straße 4
12489 Berlin

Tel. +49.30.6392-2600
Fax +49.30.6392-2602

E-Mail fbh@fbh-berlin.de
Web www.fbh-berlin.de

Innovations with Microwaves and Light

Research Reports from the Ferdinand-Braun-Institut, Leibniz-Institut für Höchstfrequenztechnik

Preface of the Editors

Research-based ideas, developments, and concepts are the basis of scientific progress and competitiveness, expanding human knowledge and being expressed technologically as inventions. The resulting innovative products and services eventually find their way into public life.

Accordingly, the *"Research Reports from the Ferdinand-Braun-Institut, Leibniz-Institut für Höchstfrequenztechnik"* series compile the institute's latest research and developments. We would like to make our results broadly accessible and to stimulate further discussions, not least to enable as many of our developments as possible to enhance everyday life.

High-power broad-area diode lasers are key components of various state-of-the-art laser systems. Accordingly, a strong research focus is on improving output power and beam quality (i.e. brightness). In this thesis, optical, electronic and thermal models are presented which allow a numerical simulation of the complicated spatio-temporal behavior of these lasers. The simulation tool developed is used to elucidate the root causes limiting laser brightness and to suggest new cavity designs improving the beam quality.

We wish you an informative and inspiring reading

Prof. Dr. Günther Tränkle
Director

Prof. Dr.-Ing. Wolfgang Heinrich
Deputy Director

The Ferdinand-Braun-Institut

The Ferdinand-Braun-Institut researches electronic and optical components, modules and systems based on compound semiconductors. These devices are key enablers that address the needs of today's society in fields like communications, energy, health and mobility. Specifically, FBH develops light sources from the visible to the ultra-violet spectral range: high-power diode lasers with excellent beam quality, UV light sources and hybrid laser systems. Applications range from medical technology, high-precision metrology and sensors to optical communications in space and integrated quantum technology. In the field of microwaves, FBH develops high-efficiency multi-functional power amplifiers and millimeter wave frontends targeting energy-efficient mobile communications as well as car safety systems. In addition, compact atmospheric microwave plasma sources that operate with economic low-voltage drivers are fabricated for use in a variety of applications, such as the treatment of skin diseases.

The FBH is a competence center for III-V compound semiconductors and has a strong international reputation. FBH competence covers the full range of capabilities, from design to fabrication to device characterization.

In close cooperation with industry, its research results lead to cutting-edge products. The institute also successfully turns innovative product ideas into spin-off companies. Thus, working in strategic partnerships with industry, FBH assures Germany's technological excellence in microwave and optoelectronic research.

Analysis of Spatio-Temporal Phenomena in High-Brightness Diode Lasers using Numerical Simulations

Dissertation
zur Erlangung des akademischen Grades

doctor rerum naturalium
— Dr. rer. nat. —

im Fach Physik
Spezialisierung: Theoretische Physik

eingereicht an der

Mathematisch-Naturwissenschaftlichen Fakultät
der Humboldt-Universität zu Berlin

von

Anissa Zeghuzi, M. Sc.

Präsidentin der Humboldt-Universität zu Berlin:
Prof. Dr.-Ing. Dr. Sabine Kunst

Dekan der Mathematisch-Naturwissenschaftlichen Fakultät:
Prof. Dr. Elmar Kulke

Gutachter:

1. Prof. Dr. Kurt Busch
2. Prof. Dr. Günther Tränkle
3. Prof. Dr. Bernd Witzigmann

Tag der mündlichen Prüfung: 28.09.2020

Bibliografische Information der Deutschen Nationalbibliothek
Die Deutsche Nationalbibliothek verzeichnet diese Publikation in der Deutschen Nationalbibliografie; detaillierte bibliografische Daten sind im Internet über http://dnb.d-nb.de abrufbar.

1. Aufl. - Göttingen: Cuvillier, 2020
Zugl.: Berlin, Univ., Diss., 2020

Nonnenstieg 8, 37075 Göttingen
Telefon: 0551-54724-0
Telefax: 0551-54724-21
www.cuvillier.de

1. Auflage, 2020
Gedruckt auf umweltfreundlichem, säurefreiem Papier aus nachhaltiger Forstwirtschaft.

ISBN 978-3-7369-7289-6
eISBN 978-3-7369-6289-7

Abstract

Broad-area lasers are edge-emitting semiconductor lasers with a wide lateral emission aperture, that enables high output powers, but also diminishes the lateral beam quality and results in their inherently non-stationary behavior. Research in the area is driven by application and the main objective is to increase the brightness, that includes both the output power and lateral beam quality. To understand the underlying spatio-temporal phenomena and to apply this knowledge in order to reduce costs for brightness optimization, a self-consistent simulation tool taking into account all essential processes is vital.

Firstly, in this work a quasi-three-dimensional opto-electronic and thermal model is presented, that describes well essential qualitative characteristics of real devices. Time-dependent traveling-wave equations are utilized to describe the inherently non-stationary optical fields, which are coupled to dynamic rate equations for the excess carriers in the active region. This model is extended by an injection current density model to accurately include lateral current spreading and spatial hole burning. Furthermore a temperature model is presented that includes short-time local heating near the active region as well as the formation of a stationary temperature profile. The former is significant for short pulse operation with high injection currents, as the fast-growing thermally induced waveguide can result in a transition from a gain-guided to an index-guided structure. Under continuous-wave operation the latter longitudinally varying stationary temperature profile leads to a near-field shrinkage at the anti-reflection coated facet, resulting in an unfavorable carrier accumulation at the stripe edges and an enhanced power density that lowers the threshold for facet damage.

Secondly, the reasons of brightness degradation, i.e. the origins of power saturation and the spatially modulated field profile are investigated. Under continuous-wave operation power saturation is mainly attributed to device heating, whereas under pulsed operation it is identified to be partly caused by spatial hole burning, lateral current spreading and two-photon absorption. Furthermore, spatio-temporal power variations play a role in the power saturation process and should not be neglected. The multi-peaked field profile of broad-area lasers is sometimes attributed to a modulation instability arising from spatial hole burning. Here, the optical field is considered to spontaneously break-up into small filaments, because the refractive index in areas of high intensity rises, thus creating a local waveguide. However, in this theory changes of the optical gain due to carrier density fluctuations and the non-stationarity of the system are not taken into account, and thus the traveling-wave simulations could not support the theory of filamentation by this indirect Kerr-type non-linearity. An alternative understanding of the transverse instabilities bases on the simultaneous lasing of a large number of lateral waveguide modes. And indeed, a post-processing analysis of the optical field obtained from a traveling-wave simulation, which makes no pre-assumption regarding modes, indicates, that in lasers with a lateral waveguide a clear mode structure is visible which is neither destroyed by the dynamics nor by longitudinal effects.

And lastly, designs that mitigate those effects that limit the lateral brightness under pulsed and continuous-wave operation are discussed. An increased beam quality can be obtained by implantation of the contact layer next to the injection stripe to counteract current spreading or by a lowered p-layer resistivity to reduced spatial hole burning. The narrowed near-field intensity resulting from the longitudinally varying temperature distribution under continuous-wave operation can be counteracted by index-guiding trenches to increase efficiency and the threshold for facet damage. And finally, a novel "chessboard laser" design is presented that utilizes longitudinal-lateral gain-loss modulation and an additional phase tailoring to theoretically provide a single-lobed far field with a full lateral far-field angle of $\Theta_{40\%} = 0.4^\circ$ (40% power content) at an injection current of 100 A under pulsed operation.

Kurzfassung

Breitstreifenlaser haben eine breite Emissionsapertur, die es ermöglicht eine hohe Ausgangsleistung zu erreichen. Gleichzeitig führt sie jedoch zu einer Verringerung der lateralen Strahlqualität und zu ihrem nicht-stationären Verhalten. Forschung in diesem Gebiet ist anwendungsgetrieben und somit ist das Hauptziel eine Erhöhung der Brillanz, die sowohl die Ausgangsleistung als auch die laterale Strahlqualität beinhaltet. Um die zugrunde liegenden raumzeitlichen Phänomene zu verstehen und dieses Wissen zu nutzen, um die Kosten der Brillanzoptimierung zu minimieren, ist ein selbst-konsistentes Simulationstool entscheidend, das alle wichtigen Prozesse beinhaltet.

Zunächst wird in dieser Arbeit ein quasi-dreidimensionales elektro-optisch-thermisches Model präsentiert, das wesentliche qualitative Eigenschaften von realen Bauteilen gut beschreibt. Zeitabhängige Wanderwellen-Gleichungen werden genutzt, um die inhärent nicht-stationären optischen Felder zu beschreiben, welche an eine Ratengleichung für die Überschussladungsträger in der aktiven Zone gekoppelt sind. Das Model wird um eine Injektionsstromdichte erweitert, die laterale Stromspreizung und räumliches Lochbrennen korrekt beschreibt. Des Weiteren wird ein Temperaturmodel präsentiert, das kurzzeitige lokale Aufheizungen in der Nähe der aktiven Zone und die Formierung einer stationären Temperaturverteilung beinhalten. Ersteres ist für Kurzpuls-Anregung mit hohen Stromdichten bedeutend, da der schnell anwachsende thermisch induzierte Wellenleiter zu einem Übergang von einer gewinn- zu einer indexgeführten Struktur führen kann. Bei Dauerstrich-Betrieb führt die letztere longitudinal variierende stationäre Temperaturverteilung zu einer Nahfeldeinschnürung an der antireflektionsbeschichteten Facette und diese wiederum zu einer unerwünschten Ladungsträgerakkumulation am Rande des Injektionsstreifens und zu einer erhöhten Leistungsdichte die zu Facettenschäden führen kann.

Im zweiten Teil werden die Gründe von Brillanzdegradierung, also die Ursprünge der Leistungssättigung und des nicht diffraktionslimitierten Fernfeldes, untersucht. Unter Dauerstrichbetrieb ist Leistungssättigung vorrangig thermisch induziert, wobei bei gepulst betriebenen Dioden als Gründe räumliches Lochbrennen, laterale Stromspreizung und Zwei-Photonen-Absorption identifiziert werden. Des Weiteren wird gezeigt, dass raumzeitliche Leistungsvariationen eine Rolle bei der Leistungssättigung spielen und nicht vernachlässigt werden sollten. Die modulierten Feldprofile von Breitstreifenlasern werden manchmal auf Modulationsinstabilitäten zurückgeführt, die durch räumliches Lochbrennen entstehen. Hier wird angenommen, dass sich das optische Feld spontan in einzelne Filamente aufgliedert, da eine Brechungsindexerhöhung in Bereichen hoher Intensität zu einem lokalen Wellenleiter führt. Da in der beschriebenen Theorie Änderungen des optischen Gewinns durch räumliche Ladungsträgeränderungen und die Dynamik des Systems nicht berücksichtig wurden, können Simulationen auf Basis der Wanderwellen-Gleichung diese Hypothese von Filamentierung nicht bestätigen. Eine alternative Erklärung der transversalen Instabilitäten gründet auf der Annahme der simultanen Anregung einer großen Anzahl lateraler Wellenleitermoden. Eine Analyse des optischen Feldes, simuliert mit dem Wanderwellenmodel, zeigt eine klare Modenstruktur in Lasern mit lateraler Wellenführung, die weder durch die Dynamik noch durch longitudinale Effekte zerstört wird.

Im letzten Teil werden Laserentwürfe besprochen, welche die laterale Brillanz verbessern. Eine erhöhte Strahlqualität kann durch Implantation der Kontaktschicht neben dem Injektionsstreifen zur Verringerung von Stromspreizung oder durch eine Verringerung der Resistivität der p-Schicht zur Eindämmung von räumlichen Lochbrennen erzielt werden. Die verengte Nahfeldintensität, die aus der longitudinalen Variation der Temperaturverteilung bei Dauerstrichbetrieb resultiert, kann durch indexführende Gräben entgegengewirkt werden, um die Effizienz zu erhöhen und Facettenschäden vorzubeugen. Zuletzt wird ein neuartiges "Schachbrettlaser" Design präsentiert, bei dem longitudinal-laterale Gewinn-Verlust-Modulation mit zusätzlicher Phasenanpassung ausgenutzt wird, um ein simuliertes Fernfeld mit einem lateralen Fernfeldwinkel von $\Theta_{40\%} = 0.4^\circ$ bei einem gepulsten Injektionsstrom von 100 A zu erhalten.

Parts of this work have been presented in

Book chapter

H. Wenzel and **A. Zeghuzi**, "High-Power Lasers," in *Handbook of Optoelectronic Device Modeling & Simulation* (J. Piprek., ed.), ch. 33, pp. 15–58, CRC Press, Taylor & Francis Group, 1rst ed., 2017.

Journals

A. Zeghuzi, H.-J. Wünsche, H. Wenzel, M. Radziunas, J. Fuhrmann, A. Klehr, U. Bandelow, A. Knigge, "Time-dependent simulation of thermal lensing in high-power broad-area semiconductor lasers", *IEEE. J. Sel. Top. Quantum Electron.*, vol. 25, no. 6, p. 1502310, 2019.

A. Zeghuzi, M. Radziunas, H. J. Wünsche, J.-P. Koester, H. Wenzel, U. Bandelow and A. Knigge, "Traveling wave analysis of non-thermal far-field blooming in high-power broad-area lasers", *IEEE J. Quantum Electron.*, vol. 55, no. 2, p. 2000207, 2019.

A. Zeghuzi, M. Radziunas, H. J. Wünsche, A. Klehr, H. Wenzel, and A. Knigge, "Influence of nonlinear effects on the characteristics of pulsed high-power broad-area distributed Bragg reflector lasers", *Opt. Quantum Electron.*, vol. 50, no. 88, pp. 1-12, 2018.

M. Radziunas, J. Fuhrmann, **A. Zeghuzi**, H.-J. Wünsche, Th. Koprucki, C. Brée, H. Wenzel, and U. Bandelow, "Efficient coupling of dynamic electro-optical and heat-transport models for high-power broad-area semiconductor lasers," *Opt. Quantum Electron.*, vol. 51, no. 69, pp. 1-10, 2019.

A. Knigge, A. Klehr, H. Wenzel, **A. Zeghuzi**, J. Fricke, A. Maaßdorf, A. Liero, and G. Tränkle "Wavelength-stabilized high-pulse-power laser diodes for automotive LiDAR," Phys. Status Solidi A, vol. 215, no. 8, p. 1700439, 2018.

M. Radziunas, **A. Zeghuzi**, J. Fuhrmann, Th. Koprucki, H.-J. Wünsche, H. Wenzel, and U. Bandelow, "Efficient coupling of the inhomogeneous current spreading model to the dynamic electro-optical solver for broad-area edge-emitting semiconductor devices," *Opt. Quantum Electron.*, vol. 49, no. 10, pp. 1–8, 2017.

International conference proceedings

A. Zeghuzi, H.-J. Wünsche, H. Wenzel, J. P. Koester, M. Radziunas, U. Bandelow, and A. Knigge, "Traveling-wave simulations of broad-area lasers", *Proc. HPDLS*, 2019.

A. Zeghuzi, H. Wenzel, H.-J. Wünsche, M. Radziunas, U. Bandelow, and A. Knigge, "Modeling of current spreading in high-power broad-area lasers and its impact on the lateral far-field divergence," *Proc. SPIE*, vol. 10526, p. 105261H, 2018.

A. Zeghuzi, M. Radziunas, A. Klehr, H.-J. Wünsche, H. Wenzel, and A. Knigge, "Influence of nonlinear effects on the characteristics of pulsed high-power BA DBR lasers," *Proc. NUSOD*, pp. 233-234, 2017.

U. Bandelow, M. Radziunas, **A. Zeghuzi**, H.-J. Wünsche, and H. Wenzel, "Dynamics in high-power diode lasers," *Proc. SPIE*, vol., p. 11356, 113560W, 2020.

A. Klehr, A. Liero, H. Wenzel, **A. Zeghuzi**, J. Fricke, R. Staske, A. Knigge, "Pico- and nanosecond investigations of the lateral nearfield of broad area lasers under pulsed high-current excitation," *Proc. SPIE*, vol. 10553, p. 105530K, 2018.

M. Radziunas, J. Fuhrmann, **A. Zeghuzi**, H.-J. Wünsche, T. Koprucki, H. Wenzel, and U. Bandelow, "Efficient coupling of heat-flow and electro-optical models for simulation of dynamics in high-power broad-area semiconductor lasers", *Proc. NUSOD*, pp. 91–92, 2018.

M. Radziunas, **A. Zeghuzi**, J. Fuhrmann, and T. Koprucki, "Efficient coupling of inhomogeneous current spreading and electro-optical models for simulation of dynamics in broad-area semiconductor lasers," *Proc. NUSOD*, pp. 231–232, 2017.

Patent application

A. Zeghuzi, J.-P. Koester, H. Wenzel, H. Christopher, and A. Knigge, DE 10 2020 108941.4 "Diodenlaser mit verringerter Strahldivergenz", 2020.

J.-P. Koester, **A. Zeghuzi**, H. Wenzel, and A. Knigge, DE 10 2020 119715.2 "Laserdiode mit räumlich selektiver Indexführung", 2020.

Acknowledgements

This work was conducted in the framework of the interdisciplinary collaboration between Ferdinand-Braun-Institut, Leibniz-Institut für Höchstfrequenztechnik (FBH) and Weierstraß-Institut für Angewandte Analysis und Stochastik (WIAS) in the scope of the *EffiLas/HoTLas* (contract 13N14005) and *EffiLAS/PLuS* (contract 13N14026) projects, supported by the German Federal Ministry of Education and Research (BMBF). This close cooperation provided not only access to excellent experimental results and scientific know-how, but also to the necessary numerical skills and infrastructure. In this way many people contributed to this work in different ways and I would like to thank all of them.

I am grateful to Prof. Dr. Kurt Busch for dedicating his time to review this work. To Prof. Dr. Günther Tränkle I would like to express my sincere gratitude for the opportunity to carry out this research in his institute, the constant interest in the progress of the work and for reviewing it. I'm grateful to Prof. Dr. Bernd Witzigmann for taking the time to be external reviewer and to Prof. Dr. Christoph Koch for chairing the dissertation committee. My sincere thanks also goes to Priv.-Doz. Dr. Uwe Bandelow for taking part in the referee committee and for fostering the close collaboration between WIAS and FBH, that has been a cornerstone of this work.

Most indebted I am to my direct supervisors Dr. Hans Wenzel and Dr. Hans-Jürgen Wünsche. This work would not have been possible without their support and advise. Their incredible knowledge and passion were immensely helpful and motivating for conducting my work. Thank you both so much for your restless and unbroken support!

I am indebted to Dr. Mindaugas Radziunas who skillfully conducted the numerical modeling and implementation of the governing equations. Thank you so much for patiently answering all my questions and for taking care of all the simulation related problems. Furthermore I would like to thank Dr. Jürgen Fuhrmann for further numerical modeling and implementation.

I am very grateful to all other colleagues from FBH, who made working here a wonderful experience. I want to thank Dr. Andrea Knigge for her constant interest in my work, her helpful remarks, and supervision. I am sincerely grateful to all the people that were involved in measurements, processing, expitaxy and mounting of the devices, such as Heike Christopher, Dr. Andreas Klehr, Jaqueline Hopp, Max Beier, Dr. Jörg Fricke, and of course many more.

I would like to especially thank Jan-Philipp Koester, who could always instantly provide the right literature for all problems and over countless lunch breaks, discussions, coffees, and almond croissants has become a really good friend.

Furthermore I am immensely grateful to my parents, partner and friends for their constant support and encouragement.

Contents

Chapter 1

Introduction and Background

High-power broad-area lasers

High-power broad-area (BA) lasers are edge-emitting semiconductor lasers with a lateral emission aperture of some tens to hundreds of micrometers which is wide compared to the emitting near infrared wavelength, making them the most efficient tool for conversion of electrical into optical energy. Due to their very high output power, high efficiency, small size and low cost in mass production there is a strong industry demand. Although they are mainly used as pump sources, their fields of application have diversified and they can be for example also employed for direct material processing or light detection and ranging (LiDAR) systems needed for autonomous driving.

For optical gain to arise in semiconductor lasers, population inversion is obtained by electrical pumping. To restrict the carrier flow and to obtain a large carrier density in the active region heterostructures with different band gap materials grown by metalorganic vapour-phase or molecular-beam epitaxy on a crystalline substrate are utilized. The broad-area lasers investigated in this thesis consist of n-doped $Al_xGa_{1-x}As$ confinement and cladding layers with appropriate Al mol fractions x, single InGaAs quantum-well (QW) active regions, p-doped confinement and cladding layers and a highly p-doped GaAs contact layer grown on a GaAs substrate, Fig. 1.1. The epitaxial structure is metalized with a gold contact and soldered p-side down on a CuW submount, Fig. 1.1(d). In high-power lasers large optical cavities are employed with a comparatively weak vertical waveguide to reduce facet load so that high output powers can be achieved even under continuous wave (CW) operation.

BA lasers reach extremely high output powers, but exhibit a complex non-stationary spatio-temporal and highly non-linear behavior as a result of the interaction of optical, electrical and thermal phenomena. Due to the interplay of spatial depletion of carriers by stimulated emission and the insufficiently fast transport of injected carriers into the depleted regions spatial hole burning occurs. On the one hand, the resulting increase of the real part of the refractive index in those regions creates a local waveguide and leads to self-focusing, which is sometimes referred to as "filamentation" [1, 2]. On the other hand, due to high stimulated recombination and reduced amplification, the carrier density and thus the optical gain is decreased in the created waveguide core, which leads to self-defocusing. The result is a highly dynamic optical field, that can be well described in the mode picture: Single mode emission becomes unstable just above threshold because, due to lateral spatial hole burning, any mode saturates the optical gain in those parts of the active layer where the mode intensity is high. The

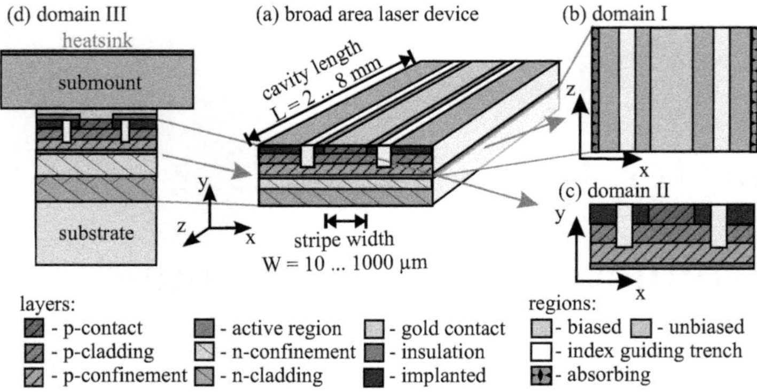

Figure 1.1: (a) Schematic representation of a BA laser and simulation domain of the (b) time-dependent traveling-wave and lateral carrier-diffusion equation, (c) inhomogeneous current-spreading model, and (d) heat transport equation. For lateral optical and current confinement index-guiding trenches can be etched. The electrical conductivity of the p-contact layer next to the injection stripe (opening of the insulation layer) is often reduced by ion implantation.

gain in other parts rises with current, bringing more modes to threshold which can be additionally supported by a built-in or thermally induced waveguide. In Zeghuzi et al., *IEEE J. Quantum Electron.*, 55(2):2000207, 2019 [3] the mode picture is supported by a theoretical analysis of the emitted laser field. The findings of [3] are discussed in more detail in Chapter 6.

A consequence of the highly dynamic field are fluctuating heat sources and a resulting non-stationary temperature profile. The time-dependent approach to thermal waveguiding and the resulting substantial refractive index changes and their significant impact on the optical field are presented in Zeghuzi et al., *IEEE J. Sel. Top. Quantum Electron.*, 25(6):1502310, 2019 [4] and discussed in more detail in Chapter 6.4.

Due to the broad emission aperture and the excitation of several lateral modes with higher near-field widths and far-field angles, BA lasers suffer from a bad lateral beam quality. It exceeds the diffraction limit already slightly above threshold, which limits the minimum achievable spot size of the emitted light. A measure of the lateral beam quality is the lateral beam parameter product $\mathrm{BPP_{lat}}$ or the beam quality factor M^2_{lat} which relates the beam quality to the beam quality of a Gaussian beam (for which $\mathrm{BPP_{lat}} = \lambda_0/\pi$),

$$\mathrm{BPP_{lat}} = \frac{w_{0,\mathrm{lat}}}{2} \cdot \frac{\Theta_{0,\mathrm{lat}}}{2} = M^2_{\mathrm{lat}} \cdot \lambda_0/\pi \tag{1.1}$$

with the lateral near-field width $w_{0,\mathrm{lat}}$ and far-field angle $\Theta_{0,\mathrm{lat}}$ (see Fig. 1.2), and lasing wavelength λ_0. Thus, the BA laser community is facing the optimization problem of achieving high output powers P_{out} and a good beam quality $\mathrm{BPP_{lat}}$ at the same time, which are combined in the target figure of merit brightness

$$B_{\mathrm{lat}} = P_{\mathrm{out}}/\mathrm{BPP_{lat}}, \tag{1.2}$$

that needs to be maximized.

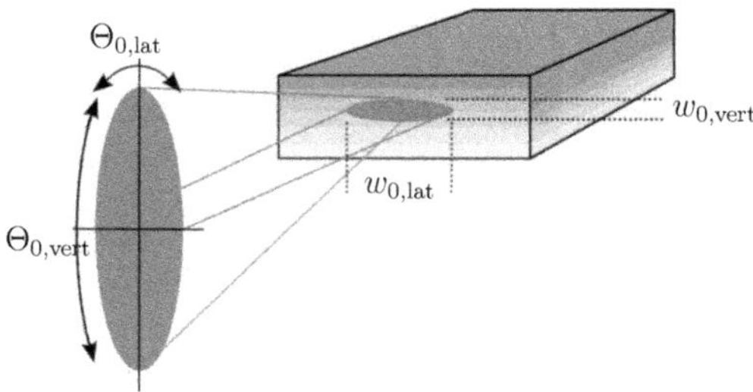

Figure 1.2: Sketch of a BA laser with emission characteristics. $w_{0,\mathrm{lat}}$ and $w_{0,\mathrm{vert}}$ denote the lateral and vertical near-field width and $\Theta_{0,\mathrm{lat}}$ and $\Theta_{0,\mathrm{vert}}$ the lateral and vertical far-field angle. Due to their broad emission stripe BA lasers suffer from a bad lateral beam quality.

With increasing current the slope of the power-current characteristics decreases due to power saturation. Under CW operation power saturation is mainly attributed to device heating as loss mechanism such as recombination, free carrier absorption and leakage currents are enhanced [5], whereas under pulsed operation, when nanosecond-current pulses with low repetition rates are applied, thermally induced power rollover is believed to be negligible. In this case non-thermal causes for power saturation are vertical carrier leakage and enhanced free-carrier absorption in the waveguide [6], longitudinal [7] and lateral [8, 9] spatial hole burning, two-photon absorption [10, 11], and mechanisms leading to gain compression [12][13], such as spectral hole burning and carrier heating as a result of finite intra-band relaxation times [14]. An analysis of the different non-thermal power-saturation effects has been published in Zeghuzi et al., *Opt. Quantum Electron.*, 50:88(1–12), 2018 [15] and is discussed in Chapter 5.

A very prominent effect that decreases the brightness is the broadening of the lateral far field with rising current ("far-field blooming"). Under CW operation it mostly results from the formation of a thermally induced waveguide, i.e. a substantial increase of the refractive index in the hot center below the contact stripe, commonly referred to as "thermal lensing". This effect is well studied, both experimentally [16, 17] and theoretically [18, 19, 20], for CW operation assuming a stationary temperature distribution. Under pulsed operation it is usually neglected because thermal build-up times of up to milliseconds are much longer than the pulse lengths. However, the heat is generated near the active layer in the same region where the guided wave is localized. This region is small and its thermal build-up time is much shorter than that of the whole device. Accordingly, under pulsed operation short-time local heating is expected to influence the optical pulse formation although time-averaged heating is negligible. Results supporting this claim, have been published in [4] and are treated in Section 6.4.1.

Also non-thermal effects play a role in the broadening of the far field [21]. Non-thermal far-field blooming is a result of current spreading [21], lateral carrier diffusion and accumulation [22], and longitudinal and lateral spatial hole burning. Current spreading and lateral carrier diffusion and accumulation not only lead to an increased optical gain at the device edges but also modify the profile of the refractive index. Ultimately in both cases a larger number of higher-order lateral modes are excited.

Semiconductor laser simulation models

The self-consistent simulation of the optical laser field, carrier transport and temperature is necessary to understand the spatio-temporal phenomena in BA lasers and to apply this knowledge in order to reduce costs for device optimization.

Due to the transverse instabilities present in BA lasers and the resulting non-stationary spatio-temporal behavior a time-dependent simulation model should be employed. Such a model needs to be sufficiently precise, whereas at the same time a good numerical performance is necessary to simulate real devices. Besides the highly dynamic and non-linear behavior, a main challenge for constructing such a model are the different time and length scales involved in the lasing process. The optical field, carrier density and temperature vary on time-scales of picoseconds, nanoseconds and microseconds, respectively, and the spatial scales of QW active region, epitaxial layers and lateral waveguide, as well as the length of the cavity range from nanometers, over micrometers to millimeters. Therefore, until now no simulation tool exists that covers all spatio-temporal scales and physical phenomena important for the description of BA lasers.

From the preceding observations it should be clear, that not all processes can be simulated on the same spatial and temporal grid and that approximations have to be made to simulate real devices within a reasonable time frame.

Generally the vertical epitaxial structure is designed to guide only a single mode. Its vertical profile remains mostly unchanged during the lasing process, so that the vertical dimension can be represented by a set of effective model parameters. Due to the nearly planar geometry of edge emitting lasers and their narrow-banded optical spectrum, in this thesis the optical field is represented by its transverse-electric x-component in the semiclassical framework of slowly-varying envelope, rotating-wave, scalar and paraxial approximations. The spatio-temporal evolution of the forward- and backward-traveling amplitudes is described by traveling-wave equations solved in the longitudinal-lateral (x, z)-plane displayed in Fig. 1.1(b) [23, 24].

The traveling-wave equations have to be coupled to the carrier reservoir via stimulated recombination and the optical field in turn is influenced by carrier density induced refractive index changes. For the description of the excess carriers in the active layer a lateral diffusion equation is solved. The injection current density entering this diffusion equation is gained self-consistently by solving the Laplace equation [18, 25] for the quasi-Fermi potential of the holes in the p-doped region sketched in Fig. 1.1(c). This model properly describes spatial hole burning, because the self-distribution of the current density is considered, as well as lateral current spreading. It is published in Zeghuzi et al., *Proc. SPIE*, 10526:105261H, 2018 [26] and discussed in Chapter 3 in more detail.

To self-consistently describe heating effects, the heat sources entering the heat-flow equation have to be calculated from the field intensity, carrier density and injection current density distributions. The resulting temperature influences the optical field via refractive index changes and temperature-dependent model parameters. However, the non-stationary heat-flow equation can't be solved on the same time and length scales as the electro-optical models. Thus in this thesis the rate of heat generation is decomposed into a time-constant mean contribution and a time-dependent fluctuating deviation from this mean value. For the treatment of pulsed operation inherent traits of heat generation and conduction can be exploited so that the temperature-induced refractive index can be self-consistently derived from the heat sources for short pulse lengths. For CW operation the electro-optical models are interatively coupled to the heat transport equations solved in the domain represented in Fig. 1.1(d). The heat model is published in [4] and discussed in more detail in Chapter 4.

The resulting time-dependent quasi-three-dimensional opto-electronic and thermal model describes well essential qualitative characteristics of real devices such as the multi-peaked, dynamic near-field structure, power-current characteristics, the time-averaged multi-peaked and not diffraction-limited near- and far-field intensities, as well as laser-output spectra.

Some laser simulation tools solve the full drift-diffusion equations to treat the current flow in the device and its interaction with the non-equilibrium carrier densities [21, 19]. These models are based on a stationary approximation and use for the description of the optical field either an expansion into linear waveguide modes [21] or a beam propagation method [19]. However, due to their inherently non-stationary and highly non-linear behavior the applicability of both methods to BA lasers is questionable. Indeed, these models do not converge at high optical output powers [27], so that spatio-temporal effects can be studied only slightly above threshold.

Due to computer restrictions, until now simulation tools based on the bidirectional traveling-wave model for the optical field [28, 29, 30] do not solve the time-dependent drift-diffusion equations, but only a lateral diffusion equation for the excess carriers in the active layer with a spatially constant injection current density below the contact stripe as source term. Such a constant-injection-current-density model, which is even used by stationary simulation tools [31, 32] oversimplifies the current flow and carrier transport in the device, because current spreading and current self-distribution are not included [33].

Theoretically, a time-dependent temperature has been considered in early models as presented in e.g. [34, 35], where they concentrated on a sophisticated microscopic description of the processes in the active layer. However, besides requiring enormous computational resources even for nanosecond transients, outer parts of devices are disregarded although a considerable portion of the heat is generated here. Heat flow was replaced by a simple local relaxation of temperature towards an ambient temperature. These features limit an application in device design.

The model proposed in this thesis is based on an existing model for the time-dependent traveling-wave equation [23]. Together with corresponding equations for the macroscopic polarization density to describe dispersion and a lateral diffusion equation for the excess carriers in the active region with a constant injection current density, this model has already been successfully applied to the simulation of a large variety of high-power laser structures [36, 37, 38, 39, 40, 41]. The improvement of the model [23] is a central part of this thesis, whereas the numerical modeling and implementation was performed by M. Radziunas and J. Fuhrmann at the Weierstrass Institute for Applied Analysis and Stochastics (WIAS). Details on the numerical modeling and used schemes can be found in [23, 42, 43, 44] and in Appendix D.

Objectives and structure of this work

The subject of this work is the analysis of spatio-temporal phenomena in BA lasers and the mitigation of effects that limit their lateral brightness. To this end a simulation model that is sufficiently precise and yet numerical applicable is derived and applied for lateral brightness optimization for pulsed and CW operation.

The following three chapters describe the time-dependent quasi-three-dimensional opto-electronic and thermal model. In Chapter 2 the parabolic paraxial wave-equation is presented that bases on the slowly-varying amplitude, rotating-wave and effective-index approximations, taking into account gain dispersion, spontaneous emission, periodic corrugations of the refractive index (i.e. Bragg gratings) and a third-order nonlinear susceptibility. In Chapter 3 the basic drift-diffusion equations will be presented. They will be reduced to an effective diffusion equation in the active region with a carrier density dependent diffusion coefficient. An advanced model for the injection current density is derived that adequately describes current self-distribution and spreading in the p-doped layers. In Chapter 4 the energy-transport model which bases on [45, 46] is summarized, paying particular attention to a consistent formulation with the model for the optical field in Chapter 2. A temperature model is presented that can self-consistently derive short-time local heating near the active region as well as the formation of a stationary temperature profil by an iterative coupling of the electro-optical models to the heat transport equations.

Chapter 5 deals with non-thermal power saturation that limits the achievable output power and as a result the brightness at high injection currents under pulsed operation. Power saturation resulting from spatial hole burning, current spreading and two-photon absorption and the impact of spatio-temporal fluctuations on the output power are discussed whereas additional effects can be included in the theoretical model by a gain compression term. The simulation results are compared to experiments.

In Chapter 6 phenomena that influence the lateral field distribution in BA lasers are examined. In the first part two mechanisms that are generally referred to when it comes to the understanding of the multi-peaked lateral field profile are discussed. On the one hand the Bespalov Talanov modulation instability [47] is investigated that describes the spontaneous break-up of the optical field into small filaments. On the other hand the decomposition of the field retrieved from the traveling wave equations into lateral waveguide modes is described. This procedure will then be used in the following sections for the investigation of the lasing process with regards to the mode picture. In the second part of Chapter 6 the non-thermal broadening of the far field with current and its dependence on series resistivity, sheet resistance, mobility, as well as carrier induced refractive index changes is investigated. These effects are discussed in the mode picture and possibilities for beam quality improvement are derived. In the last part of Chapter 6 the influence of slow and fast contributions to the time-dependent temperature on the lateral near- and far-field distributions are investigated. On the one hand, the effect of a thermally induced waveguide under short pulse operation with extremely high injection currents is exemplary investigated for laser operation with 10 ns long pulses. On the other hand the time-dependent approach to CW operation is discussed. In particular the strong spatio-temporal sub-nanosecond fluctuations of the heat sources on wave propagation are examined. Furthermore the front facet near-field narrowing as a result of the time-averaged longitudinally varying temperature profile is derived self-consistently. To counteract it, index-guiding trenches filled with an insulator can be etched next to the injection stripe. The simulation results are compared to experiments.

In Chapter 7 methods for the improvement of the lateral brightness are discussed. In the first part of Chapter 7 the brightness improvement by varying the built-in index step of index-guiding trenches, their width and distance to the injection stripe will be discussed. A further current-path tailoring by deep implantation of the laser next to the injection stripe is investigated as well. The simulation results are compared to experiments. In the second part of Chapter 7 a lateral and longitudinal structuring of the contact region is investigated. Laser arrays show a narrow single lobe around 0° far-field angle when the fields in each stripe element are co-phasal. As this is only achieved at currents near threshold, additional mode-selection mechanisms are discussed, such as a Talbot-type spatial filter where free running sections with lengths corresponding to half the Talbot length are alternately included. An additional longitudinal structuring with periods corresponding to the Talbot length and additional phase tailoring for the suppression of the out-of-phase mode are investigated. The simulation results are compared to experiments.

Chapter 2

Optical field model

To model laser operation far above threshold the electromagnetic field can be treated semiclassicaly by the Maxwell equations with a spontaneous current-density sources term [48, 49]. The full vectorial wave equation can be significantly simplified using certain properties of the emitted laser light: The investigated edge emitting devices are longitudinally more extended than laterally and the emitted laser light has a dominant propagation in z-direction. The growth direction (y-direction) defines a nearly planar geometry, so that the optical field is expressed by its transverse-electric x-component. As the transverse refractive index changes are small and the laser emits in a small frequency band around a central frequency further approximations can be made.

Accordingly a parabolic paraxial wave-equation for the slowly-varying forward- and backward-traveling field envelopes are gained as described in Section 2.1. The traveling-wave equations are coupled to dynamic equations for the macroscopic polarization density and carrier density, so that this set of equations is formally similar to the system of Maxwell-Bloch equations for a two level atom [28]. In Section 2.2 a balance equation for the radiative energy density and the associated energy flux given by the Poynting vector is derived. Supposing a well-designed vertical waveguide, the normalized fundamental vertical mode $\phi(y)$ remains unchanged during laser operation. Accordingly the vertical direction can be represented by effective model parameters. This approximation termed "effective index method" is explicitly executed in Section 2.3 to clarify the derivation of model parameters. Together with the model equations presented in Chapter 3 and 4 for the carrier density and temperature this system of equations describes well essential qualitative characteristics of real devices. The retrieval of the corresponding characteristics that can be used to compare to experimental data is described in more detail in Section 2.4.

2.1 The traveling-wave equations

The governing optical equations can be derived semiclassically from Maxwells equations using a spontaneous current-density source term. A detailed derivation can be found in [49]. Here, the important approximations and underlying properties of the emitted laser light are shortly recapitulated.

As a result of the nearly planar geometry of edge emitting semiconductor lasers defined by the epitaxial layer structure the electromagnetic field is either mainly transverse electric (TE) or transverse magnetic (TM) polarized. In this work the optical field is represented by its transverse-electric x-component. Furthermore the electric field is separated into forward- and backward-traveling wave amplitudes $E^{\pm}(\vec{r},t)$ and by choosing an appropriate reference wave vector $\bar{n}k_0$, where $\bar{n}$ is the real valued reference index and k_0 the free-space wavevector, and center frequency ω_0, the transverse-electric optical field is represented by

$$\vec{E}(\vec{r},t) = \vec{e}_x \frac{1}{2} E(\vec{r},t) \mathrm{e}^{i\omega_0 t} + \text{c.c.} \tag{2.1}$$

$$\text{with} \quad E(\vec{r},t) = E^{+}(\vec{r},t)\mathrm{e}^{-i\bar{n}k_0 z} + E^{-}(\vec{r},t)\mathrm{e}^{i\bar{n}k_0 z} \tag{2.2}$$

and $k_0 = 2\pi/\lambda_0$ with the center wavelength λ_0.

Several important properties of the emitted laser light can be used to significantly simplify the wave equation. Firstly, the wave propagation within a laser resonator has a dominant propagation direction (here z-direction). Furthermore lasers generally emit within a small frequency range around a center frequency ω_0 corresponding to the optical transitions in the active material. Thus, it can be assumed that the envelopes of the forward- or backward-traveling waves $E^{\pm}(\vec{r},t)$ vary slowly in time and space compared to the optical wavelength or spatial period defined in Eqs. (2.1) and (2.2) by the reference wave vector $\bar{n}k_0$ and center frequency ω_0.

This slowly-varying envelope approximation implies, that the second z-derivative of the electric field $|\partial_z^2 E^{\pm}|$ is small against $|2\bar{n}k_0\partial_z E^{\pm}|$ (paraxial approximation) and likewise that the second time derivative $|\partial_t^2 E^{\pm}|$ is small against $|2\omega_0\partial_t E^{\pm}|$ and that both can be neglected in the final equation. Based on the same assumption, terms that rapidly oscillate in time $\mathrm{e}^{i2\omega_0 t}$ (rotating wave approximation) and space $\mathrm{e}^{\pm i2\bar{n}k_0 z}$ can be neglected.

Furthermore, due to the small transverse refractive index change only the scalar wave equation has to be solved. Taking into account these approximation the slowly-varying envelope amplitudes of the forward- and backward-traveling waves obey the traveling-wave equation [49]

$$\begin{aligned}
\frac{1}{v_\mathrm{g}}\partial_t E^{\pm}(\vec{r},t) \pm \partial_z E^{\pm}(\vec{r},t) = &-\frac{i}{2\bar{n}k_0}\left(\partial_x^2 + \partial_y^2\right) E^{\pm}(\vec{r},t) \\
&-\frac{ik_0}{2\bar{n}}\Delta n^2(\vec{r},t) E^{\pm}(\vec{r},t) - \frac{ik_0 n_r \varepsilon_0 c}{2}\Delta n_2(\vec{r},t)\left[|E^{\pm}(r,t)|^2 + 2|E^{\mp}(r,t)|^2\right] E^{\pm}(\vec{r},t) \\
&-ik_0\frac{\eta^{\pm}(r,\omega_0)}{2\bar{n}} E^{\mp}(r,t) - \frac{n_\mathrm{r}(\vec{r},t) G_\mathrm{r}(\vec{r},t)}{2\bar{n}}\left[E^{\pm}(\vec{r},t) - \mathcal{P}^{\pm}(\vec{r},t)\right] + F_\mathrm{sp}^{\pm}(\vec{r},t).
\end{aligned} \tag{2.3}$$

Here $v_\mathrm{g} = c/n_g$ is the group velocity with the speed of light c and group index $n_g = \bar{n} + \omega_0 \partial_\omega \bar{n}(\omega)|_{\omega=\omega_0}$ and ε_0 the vacuum permittivity.

The first term on the right hand side of Eq. (2.3) accounts for transverse refraction of the optical field. $\Delta n^2(\vec{r},t)$ describes the deviation of the complex refractive index $n^2 = 1 + \chi$ to the real valued reference index $\bar{n}^2 = 1 + \bar{\chi}$, where χ and $\bar{\chi}$ are the complex and real valued reference susceptibilities, respectively,

$$\Delta n^2(\vec{r},t) \equiv n^2(\vec{r},t,\omega_0) - \bar{n}^2 = \Delta n_\mathrm{r}^2(\vec{r},t) + i\frac{n_\mathrm{r}(\vec{r},t)\left[g(\vec{r},t,\omega_0) - \alpha(\vec{r},t)\right]}{k_0}.$$

Here Δn_r^2 is the real part of Δn^2, n_r the real part of n and g and α the coefficients of optical gain and absorption due to transitions between the conduction and valence bands, respectively. The coefficient of the optical gain g couples the traveling-wave equation for the optical field to the dynamic rate equation of the carrier reservoir (3.39) via the rate of stimulative recombination (2.18). The complex refractive index Δn has contributions from carrier density and temperature as discussed in more detail in Section (2.3). The absorption coefficient also includes background absorption in the p- and n-doped layers due to the doping profile and furthermore additional scattering losses can be included. Δn_2 describes additional changes of the complex refractive index due to a third-order nonlinear susceptibility $\chi^{(3)}$, that results in a nonlinear contribution to the polarization density [1, 49],

$$\Delta n_2 \equiv \frac{3}{4 n_{\mathrm{r}} \varepsilon_0 \bar{n} c} \chi^{(3)} = n_2 - i \frac{\beta}{2 k_0}. \tag{2.4}$$

The coefficients n_2 and β are the optical Kerr coefficient and the two-photon absorption coefficient related to the time-averaged intensity $I = \varepsilon_0 \bar{n} c |E|^2/2$, respectively, where n_2 has the unit m^2/W and β the unit m/W. For the calculation of n_2 and β the model given in [50] is used, which is described in more detail in Appendix B. The optical Kerr coefficient n_2 describes a change of the real part of the refractive index depending on the intensity I and can result in self-focusing as discussed in Section 6.1.2. β describes the absorption of two-photons in the confinement and cladding layers where the energy gaps are larger than the photon energy generated by stimulated emission. Power saturation resulting from two-photon absorption is studied for pulsed laser operation in Chapter 5.1.

The term $\eta^{\pm}$ accounts for periodic corrugations of n^2 with spatial period length Λ for which the Bragg condition $2\pi/\Lambda = 2\bar{n} k_0$ holds. These refractive index corrugations are used as Bragg gratings in DFB or DBR lasers for longitudinal mode selection. The reflectivity of the Bragg grating at a certain wavelength is determined by the coupling coefficient and the length of the grating L_{DBR} and can be found in e.g. [51].

Dispersion is modelled via a Lorentzian $\mathfrak{L}(\vec{r}, \omega)$ response function in the frequency domain [28]

$$\mathcal{P}(\vec{r}, \omega) = \mathfrak{L}(\vec{r}, \omega) E(\vec{r}, \omega) \tag{2.5}$$

with

$$\mathfrak{L}(\vec{r}, \omega) = \frac{\Gamma_{\mathcal{P}}(\vec{r})}{i[\omega - \omega_0 - (\Omega_{\mathcal{P}}(\vec{r}) - \omega_0)] + \Gamma_{\mathcal{P}}(\vec{r})}. \tag{2.6}$$

G_r in Eq. (2.3) gives the height of the Lorentzian $\mathfrak{L}(\vec{r}, \omega)$ which achieves its maximum value at the frequency $\Omega_{\mathcal{P}}$. The imaginary part $\operatorname{Im} \mathfrak{L}(r, \omega)$ has a half width $\Gamma_{\mathcal{P}}$ at half of the maximum. The approximation is valid only within a small frequency region around ω_0 corresponding to the frequency range of optical transitions in the active material. Inserting (2.6) into (2.5), multiplying it with the denominator and taking the inverse Fourier transform with respect to $\omega - \omega_0$ the ordinary differential equation

$$\partial_t \mathcal{P}(\vec{r}, t) = [i(\Omega_{\mathcal{P}}(\vec{r}) - \omega_0) - \Gamma_{\mathcal{P}}(\vec{r})] \mathcal{P}(\vec{r}, t) + \Gamma_{\mathcal{P}}(\vec{r}) E(\vec{r}, t) \tag{2.7}$$

is obtained. A similar composition as for the electric field is used for the dispersive polarization density

$$\mathcal{P}(\vec{r}, t) = \mathcal{P}^+(\vec{r}, t) \mathrm{e}^{-i\bar{n}k_0 z} + \mathcal{P}^-(\vec{r}, t) \mathrm{e}^{i\bar{n}k_0 z}. \tag{2.8}$$

Inserting (2.8) and (2.2) into (2.7), multiplying with $\mathrm{e}^{\pm i\bar{n}k_0 z}$ and again neglecting $\mathrm{e}^{\pm 2i\bar{n}k_0 z}$,

$$\partial_t \mathcal{P}^{\pm}(\vec{r}, t) = i(\Omega_{\mathcal{P}}(\vec{r}) - \omega_0) \mathcal{P}^{\pm}(\vec{r}, t) + \Gamma_{\mathcal{P}}(\vec{r}) \left[E^{\pm}(\vec{r}, t) - \mathcal{P}^{\pm}(\vec{r}, t) \right] \tag{2.9}$$

is obtained, which is solved together with Eq. (2.3) in every time step.

The stochastic Langevin forces $F_{\mathrm{sp}}^{\pm}$ describe spontaneous emission. After each time step Δt the Langevin forces $\Delta F_{\mathrm{sp}}^{\pm}$ are represented by random numbers added to the fields leaving a longitudinal subsection $\Delta z = c\Delta t/n_{\mathrm{g}}$. The mean square of the random numbers added to the field is proportional to $\beta_{\mathrm{sp}} R_{\mathrm{sp}}$ [52] where R_{sp} given in Eq. (3.21) is the rate of spontaneous emission into all modes and the dimensionless factor β_{sp} the ratio between spontaneous emission rate into a guided mode and the recombination rate.

Eqs. (2.3) must be supplemented by appropriate boundary conditions. At the plane facets of the laser located at $z = 0$ and $z = L$

$$\begin{aligned} E^{+}(x, y, 0, t) - r_0 E^{-}(x, y, 0, t) &= 0 \\ E^{-}(x, y, L, t) - r_L e^{-i2\bar{n}k_0 L} E^{+}(x, y, L, t) &= 0 \end{aligned} \tag{2.10}$$

hold. In the paraxial approximation the amplitude reflection coefficients r_0 and r_L are input parameters which have to be calculated in advance. Periodic boundary conditions are assumed at the transverse boundaries of the sufficiently broad simulation domain $[-X, X]$,

$$E^{\pm}(x + 2X, y, z, t) = E^{\pm}(x, y, z, t). \tag{2.11}$$

Furthermore, just before the transverse boundaries artificial absorbing regions are introduced, see Fig. 1.1(b), which ensure that no possibly remaining field intensities are reinjected into the simulation domain through the transverse borders.

The optical power is

$$P^{\pm}(z) = \frac{\varepsilon_0 \bar{n} c}{2} \iint |E^{\pm}(x, y, z)|^2 \, dx dy. \tag{2.12}$$

Correspondingly the output power at rear ($z = 0$) and front ($z = L$) facet is given by

$$P_0 = (1 - |r_0|^2) P^{-}(0) \quad \text{and} \quad P_L = (1 - |r_L|^2) P^{+}(L), \tag{2.13}$$

respectively.

2.2 Balance of radiative energy

In this section a balance equation for the radiative energy density u_{rad} and the associated energy flux given by the Poynting vector S_{rad} is derived. The derived equation will also be of importance later in the explicit derivation of the heat source density from Eq. (4.6).

The radiative energy density u_{rad} of an electromagnetic wave for a dispersive medium can be calculated from $||E(r, t)||^2 = |E^{+}|^2 + |E^{-}|^2$ as [53]

$$u_{\mathrm{rad}} = \frac{\varepsilon_0 \bar{n} n_g}{2} ||E(\vec{r}, t)||^2. \tag{2.14}$$

Multiplying Eq. (2.3) with the complex conjugate $E^{\pm *}$, multiplying the complex conjugate of Eq. (2.3) with $E^{\pm}$ and adding all equations results in

$$\begin{aligned}\frac{1}{v_{\rm g}}\partial_t\|E\|^2 = &\frac{1}{\bar{n}k_0}{\rm Im}\left[\partial_x(E^{+*}\partial_x E^+ + E^{-*}\partial_x E^-) + \partial_y(E^{+*}\partial_y E^+ + E^{-*}\partial_y E^-)\right]\\ &+\partial_z\left[|E^-|^2 - |E^+|^2\right]\\ &+\frac{n_{\rm r}(g-\alpha)}{\bar{n}}\|E\|^2 + \frac{n_{\rm r}g_{\rm r}}{\bar{n}}\left[{\rm Re}(E^{+*}\mathcal{P}^+ + E^{-*}\mathcal{P}^-) - \|E\|^2\right]\\ &-\frac{n_{\rm r}\varepsilon_0 c}{2}\beta(\|E\|^4 + 4|E^+|^2|E^-|^2) + 2{\rm Re}(E^{+*}F^+_{\rm sp} + E^{-*}F^-_{\rm sp})\end{aligned} \tag{2.15}$$

with $\|E\|^2 = |E^+|^2 + |E^-|^2$ and $\|E\|^4 = |E^+|^4 + |E^-|^4$. Note, that the Bragg grating coupling term vanishes, because $\eta^{\pm}$ is real. Eq. (2.15) can thus be interpreted as a balance equation for the radiative energy density $u_{\rm rad}$ and the associated energy flux given by the Poynting vector $S_{\rm rad}$,

$$\begin{aligned}\partial_t u_{\rm rad} + \nabla\cdot S_{\rm rad} = &\hbar\omega_0 R_{\rm st} - \frac{\varepsilon_0 c}{2}n_{\rm r}\alpha\|E\|^2 - \left(\frac{\varepsilon_0 c}{2}\right)^2\bar{n}n_{\rm r}\beta(\|E\|^4 + 4|E^+|^2|E^-|^2)\\ &+\varepsilon_0 c\bar{n}{\rm Re}(E^{+*}F^+_{\rm sp} + E^{-*}F^-_{\rm sp}),\end{aligned} \tag{2.16}$$

with the divergence of the energy flux,

$$\begin{aligned}\nabla\cdot S_{\rm rad} = &-\frac{\varepsilon_0 c}{2k_0}{\rm Im}\left[\partial_x(E^{+*}\partial_x E^+ + E^{-*}\partial_x E^-) + \partial_y(E^{+*}\partial_y E^+ + E^{-*}\partial_y E^-)\right]\\ &-\frac{\varepsilon_0\bar{n}c}{2}\partial_z\left[|E^-|^2 - |E^+|^2\right],\end{aligned} \tag{2.17}$$

and the rate of stimulative recombination

$$R_{\rm st} = \frac{\varepsilon_0 c n_{\rm r}}{2\hbar\omega_0}\left[g\|E\|^2 + g_{\rm r}\sum_{\nu=+,-}{\rm Re}\left(E^{\nu *}\mathcal{P}^\nu - E^{\nu *}E^\nu\right)\right]. \tag{2.18}$$

The first term on the right hand side of Eq. (2.16) describes the increase of the energy due to stimulated emission, the second and third term describe the energy decrease due to absorption, including two photon absorption, and the last term describes the increase of energy due to spontaneous emission.

Integrating (2.17) over the device volume V gives the radiation leaving the cavity as $P_0 + P_L$ from (2.12) and (2.13),

$$\int_V \nabla\cdot S_{\rm rad}\, dV = P_0 + P_L. \tag{2.19}$$

Here the boundary conditions (2.10) and (2.11) have been inserted, so that the transverse energy flux density, first term in (2.17), vanishes.

2.3 Effective longitudinal-lateral projected equations

In this section the approach called "effective index method" [54] in the semiconductor laser community is used to project the basic equations onto the dominant vertical mode. Assuming a well-designed vertical waveguide, the vertical mode profile $\phi(y, x, z)$ does not vary strongly along (x, z) and remains unchanged during laser operation. Here (x, z) only indicates a parametric dependence. For every (x, z) the vertical mode $\phi(y)$ can be calculated in advance for the cold cavity, so that for the electrical field the Ansatz

$$E^{\pm}(\vec{r}, t) = \sqrt{\frac{2d\hbar\omega_0}{\varepsilon_0 \bar{n} n_{\mathrm{g}}}} \phi(y) u^{\pm}(x, z, t), \tag{2.20}$$

and similarly for the polarization density

$$\mathcal{P}^{\pm}(\vec{r}, t) = \sqrt{\frac{2d\hbar\omega_0}{\varepsilon_0 \bar{n} n_{\mathrm{g}}}} \phi(y) p^{\pm}(x, z, t), \tag{2.21}$$

is used, where the parametric dependence on (x, z) has been omitted for simplicity and the mode profile $\phi(y)$ is normalized according to $\int |\phi|^2 \, dy = 1$. d and $\hbar$ are the thickness of the active region and the Planck constant, respectively. The scaling in Eq. (2.20) is chosen such that $|u^{\pm}|^2$ is a photon density (unit m^{-3}). With Eq. (2.12) the power is given by

$$P^{\pm} = \hbar\omega_0 v_{\mathrm{g}} d \int |u^{\pm}|^2 \, dx. \tag{2.22}$$

$\phi(y)$ is a solution of a vertical waveguide equation with a real-valued index profile $n_0(y)$ not dependent on carrier densities and temperature,

$$\partial_y^2 \phi(y) + k_0^2 n_0^2(y) \phi(y) = k_0^2 n_{\mathrm{eff},0}^2 \phi(y) \tag{2.23}$$

where the real-valued effective index $n_{\mathrm{eff},0}$ is the eigenvalue of (2.23). The assumption of a carrier densities and temperature independent vertical mode profile $\phi(y)$ is for example supported in [55], where the measured vertical near-field width and far-field angle remain nearly unchanged while tuning the output power.

Inserting (2.20) into (2.3), multiplying with ϕ and integrating along y yields the equations

$$\begin{aligned} \frac{1}{v_g} \partial_t u^{\pm}(x, z, t) = &- \frac{i}{2\bar{n}k_0} \partial_x^2 u^{\pm}(x, z, t) \mp \partial_z u^{\pm}(x, z, t) \\ &- ik_0 \Delta n_{\mathrm{eff}} u^{\pm}(x, z, t) - ik_0 \Delta n_{2,\mathrm{eff}}^{\pm}(x, z, t) u^{\pm}(x, z, t) \\ -ik_0 \kappa^{\pm}(x, z) u^{\mp}(x, z, t) &- \frac{g_r(x, z) n_r}{2\bar{n}} \left(u^{\pm}(x, z, t) - p^{\pm}(x, z, t) \right) + f_{\mathrm{sp}}^{\pm}(x, z, t). \end{aligned} \tag{2.24}$$

For $u^{\pm}$ the same boundary conditions (2.10) and (2.11) as for $E^{\pm}$ apply. Δn_{eff} comprises all deviations of the complex refractive index from the reference index $\bar{n}$ and is defined as

$$\Delta n_{\mathrm{eff}} \equiv \frac{n_{\mathrm{eff}}^2(x, z, t) - \bar{n}^2}{2\bar{n}} = \Delta n_{\mathrm{eff,r}} + i \frac{\bar{n} \left(g_{\mathrm{eff}} - \alpha_{\mathrm{eff}} \right)}{k_0}, \tag{2.25}$$

where $\Delta n_{\mathrm{eff,r}}$ is the real part of Δn_{eff} and given by

$$\Delta n_{\mathrm{eff,r}} \equiv \frac{n_{\mathrm{eff},0}^2 - \bar{n}^2 + \int (n_r^2 - n_0^2) |\phi|^2 dy}{2\bar{n}}. \tag{2.26}$$

The imaginary part of Δn_{eff} gives the effective gain and absorption calculated by

$$g_{\mathrm{eff}} - \alpha_{\mathrm{eff}} = \frac{\int n_r (g - \alpha) |\phi|^2 dy}{\bar{n}}. \tag{2.27}$$

$\Delta n_{\text{eff,r}}$ comprises built-in modifications of the effective index due to optionally etched index guiding trenches Δn_0, carrier density N induced index contributions Δn_N and temperature T induced index contributions Δn_T,

$$\Delta n_{\text{eff,r}}(x,z,N) = \Delta n_0(x,z) + \Delta n_N(x,z,N) + \Delta n_T(x,z,T) \tag{2.28}$$

$$\text{with} \quad \Delta n_N = -\sqrt{n'_N N} \qquad (2.29) \qquad \text{and} \quad \Delta n_T = \int n'_T T |\phi|^2 \, dy. \qquad (2.30)$$

Here, n'_N and n'_T are the carrier density induced differential index and the temperature induced differential index, respectively.

The effective gain is modeled via

$$g_{\text{eff}} = \frac{g' \ln(N/N_{\text{tr}})}{1 + \epsilon_{\text{s}} ||u||^2} \tag{2.31}$$

with $||u||^2 = |u^+|^2 + |u^-|^2$. Here g' is the differential modal gain and N_{tr} the transparency carrier density, which are gained by a fit to results of a microscopic gain model [56] (*cf.* Appendix A). The gain compression coefficient ϵ_{s} is used to describes effects of gain compression that are not self-consistently included in the model, such as carrier heating and spectral hole burning. It is taken from [49] and multiplied with the active region confinement factor Γ, $\epsilon_{\text{s}} = \Gamma \cdot 10^{-23}\ \text{m}^3$. The denominator in the equation for the effective gain g_{eff} is sometimes also written in terms of power as $1 + P/P_{\text{s}}$ with the saturation power $P_{\text{s}} = dW v_{\text{g}} \hbar\omega / \epsilon_{\text{s}}$. For $\epsilon_{\text{s}} = \Gamma \cdot 10^{-23}\ \text{m}^3$ the saturation power is $P_{\text{s}} \approx 150$ W. A further discussion on gain compression can be found in Section 5.1.

α_{eff} comprise losses due to free carrier absorption in the cladding and waveguide layers $\alpha_{0,\text{eff}}$ and free carrier absorption in the active region $\alpha_{N,\text{fc}}$,

$$\alpha_{\text{eff}} = \alpha_{0,\text{eff}} + \alpha_{N,\text{fc}}, \tag{2.32}$$

$$\text{with} \quad \alpha_{0,\text{eff}} = \int \frac{n_r \alpha_0}{\bar{n}} |\phi|^2 dy \qquad (2.33) \qquad \text{and} \quad \alpha_{N,\text{fc}} = \Gamma f_N N, \qquad (2.34)$$

where α_0 is the background absorption coefficient according to the doping profile calculated from Eq. (A.1), where scatting losses can be included, Γ is the active region confinement factor, and f_N the cross section for free-carrier absorption in the active region.

$\Delta n_{2\text{eff}}^{\pm}$ describes effective changes of the complex refractive index due to the Kerr effect and two-photon absorption and is calculated according to

$$\begin{aligned} \Delta n_{2\text{eff}}^{\pm} &= \int \frac{n_r \Delta n_2 |\phi|^4}{\bar{n}} dy \cdot \hbar\omega_0 v_{\text{g}} d(|u^{\pm}(x,z,t)|^2 + 2|u^{\mp}(x,z,t)|^2) \\ &= \Delta n_{2\text{r,eff}}^{\pm} - i\frac{\alpha_{2\text{P,eff}}^{\pm}}{2k_0}, \end{aligned} \tag{2.35}$$

$$\text{with} \quad \alpha_{2\text{P,eff}}^{\pm} = \beta' \hbar\omega_0 v_g d(||u||^2 + |u^{\mp}|^2) \quad \text{with} \quad \beta' = \int \frac{n_r \beta}{\bar{n}} |\phi|^4 dy \tag{2.36}$$

$$\text{and} \quad \Delta n_{2\text{r,eff}}^{\pm} = n'_2 \hbar\omega_0 v_g d(||u||^2 + |u^{\mp}|^2) \quad \text{with} \quad n'_2 = \int \frac{n_r n_2}{\bar{n}} |\phi|^4 dy. \tag{2.37}$$

Here β' is the effective two-photon absorption coefficient in units of W^{-1} and n_2' the effective optical Kerr coefficient in units of mW^{-1}. Further information on the simulation parameters can be found in Appendix A.

Furthermore the Bragg grating coupling coefficient, spontaneous emission factor and stimulative recombination rate are given by

$$\kappa^{\pm}(x,z) = \int \frac{\eta^{\pm}(x,y,z,\omega_0)}{2\bar{n}} |\phi|^2 \, dy, \quad f_{\rm sp}^{\pm}(x,z,t) = \int F_{\rm sp}^{\pm}(x,y,z,\omega_0)|\phi|^2 \, dy \tag{2.38}$$

$$\text{and} \quad R_{\rm st} = v_g \left[g_{\rm eff} \|u\|^2 + g_r \sum_{\nu=+,-} \mathrm{Re}\left(u^{\nu *} p^{\nu} - u^{\nu *} u^{\nu} \right) \right]. \tag{2.39}$$

By inserting (2.20) and (2.21) into (2.9), multiplying with ϕ and integrating along y, similarly for the polarization density the differential equation

$$\partial_t p^{\pm}(x,z,t) = \gamma(x,z)(u^{\pm}(x,z,t) - p^{\pm}(x,z,t)) + i\delta\omega(x,z)p^{\pm}(x,z,t), \tag{2.40}$$

is gained, where g_r in (2.39), γ and $\delta\omega$ are the Lorentzian amplitude, the half width at half maximum and the gain peak frequency detuning from the center frequency of the Lorentzian used to model dispersion.

Eq. (2.24) together with the corresponding equation for the polarization density (2.40) and a lateral diffusion equation for the excess carriers in the active region (3.39) have already been successfully applied to the simulation of a large variety of high-power laser structure [36, 37, 38, 39, 40, 41].

2.4 Retrieval of real device characteristics

Besides equations for the polarization density (2.40) and carrier transport in the active region (3.39) in the next two chapters 3 and 4 the traveling-wave equations (2.24) will be complemented by equations for the injection current density (3.43) and (3.44) and temperature (4.45) and (4.46). Together, this system of equations describes well essential qualitative characteristics of real devices such as the multi-peaked, dynamic near-field structure, Fig. 5.4, power-current (PI) characteristics, Fig. 5.1, the time averaged multi-peaked and not diffraction-limited near and far-field intensities, Fig. 3.4, as well as spectra, Fig. 6.8(a). The retrieval of the corresponding characteristics is described here in more detail. They can also be obtained experimentally, so that when reasonable in the corresponding sections, the theoretical predictions will be compared with experimental data. In this work a particular attention is payed to the qualitative agreements to gain an insight into the underlying physical processes.

The time trace of the near-field intensity at the rear facet ($z = L$), Fig. 5.4, is given by

$$I_{\rm NF}(x,t) = \hbar\omega v_g |u^{+}(x, z = L, t)|^2. \tag{2.41}$$

From the complex near-field amplitude at the output facet $\propto u^{+}(x, z = L, t)$ under the condition that the field is evaluated at a far distance D from the facet, i.e. $w_0^2/\lambda \ll D$, the complex far field as function of the lateral far-field angle Θ can be calculated via a Fourier transformation,

$$\hat{u}_{\rm FF}(\Theta, t) \propto \int_{-\infty}^{\infty} u^{+}(x, z = L, t) \cdot e^{ik_0 \sin(\Theta) x} \, dx. \tag{2.42}$$

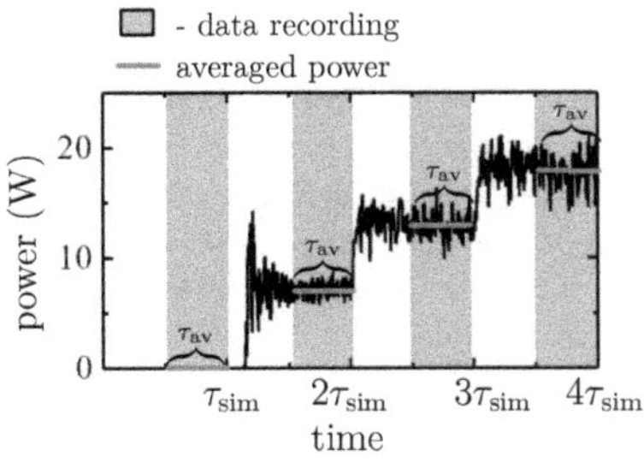

(a) time trace of the optical output power

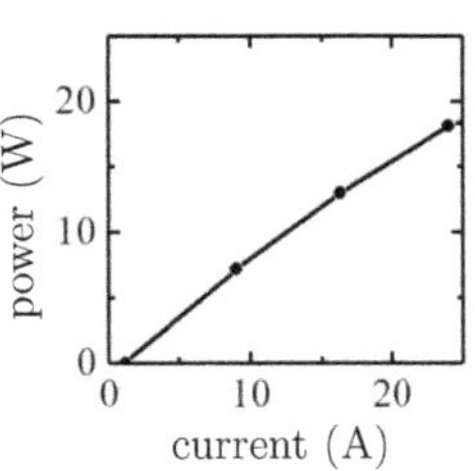

(b) power-current characteristics

Figure 2.1: Simulation of the PI characteristics with the traveling-wave model. (a) Time trace of the optical output power. The simulation time of each bias is τ_{sim}. Data is recorded for the time τ_{av} of that time span, then the voltage value is changed. (b) Power-current characteristics. Values displayed in the PI characteristics are averaged powers for the time span τ_{av} and corresponding current values for the applied bias.

The corresponding time trace of the far-field intensity is given as,

$$I_{\text{FF}}(\Theta, t) = |\hat{u}_{\text{FF}}(\Theta, t)|^2. \tag{2.43}$$

For the time-averaged near and far fields, e.g. Fig. 3.4, a bias or current value is set to a constant value and the transient simulation is performed for the time interval τ_{sim}. After the relaxation oscillation the field is recorded for the time τ_{av} during the last nanoseconds of τ_{sim}, Fig. 2.1, then the data is averaged. τ_{sim} and τ_{av} are given where the corresponding data is discussed and chosen so that the mean value does not change significantly when using much longer time spans.

Important figures of merit to assess the beam quality are the full near-field width w_0 and far-field angle Θ_0 containing 95% of the power. They are extracted from the time averaged near and far-field profiles mostly at the output facet at $z = L$.

For the calculation of the PI characteristics with the dynamic traveling-wave model, a staircase voltage ramp is used. The calculated transient of the output power

$$P_{\text{out}}(t) = (1 - |r_L|^2)P^+(z = L, t), \tag{2.44}$$

with P^+ from (2.12) and the front facet amplitude reflection coefficient r_L, is shown in Fig. 2.1(a) for the first few stairs. After the relaxation oscillations at the beginning, the system is still governed by spatio-temporal fluctuations around a mean value inherent to BA lasers. The mean value over the time interval τ_{av} of the last nanoseconds of each stair is used for the plot of the PI curves in Fig. 2.1(b).

Furthermore from the complex near fields at the output facet $u^+(x, z = L, t)$ the spatially resolved intensity distribution, as shown in Fig. 6.8(a), can be obtained by a Fourier transformation,

$$|\hat{u}_\lambda(\lambda, x)|^2 \propto \left| \int_{-\infty}^{\infty} u^+(x, z = L, t) \cdot e^{i(2\pi c/\lambda)t} \, dt \right|^2. \tag{2.45}$$

This is numerically done using a fast Fourier transform algorithm, where a Blackman window is applied to reduce frequency leakage.

2.5 Summary

The basis of the optical model is a traveling wave equation for the slowly-varying complex amplitudes given in the semiclassical framework of rotating-wave, scalar, and paraxial approximations. The traveling-wave equations are dependent on a complex refractive index that includes changes due to an build in index, carrier density, temperature and a third-order suszeptibility. Furthermore the model includes coupling to a Bragg grating and the carrier density dependent optical gain accounts for gain compression. To model gain dispersion, the traveling-wave equations are coupled to a differential equation for complex slowly-varying amplitudes of the polarization density solved in time domain, that approximates the dispersion of gain with a Lorentzian function. For a well-designed vertical waveguide, the vertical field profile is mono-modal and remains unchanged during laser operation. Under these assumptions the traveling-wave equations are reduced to effective equations that can be solved in the longitudinal-lateral (x, z)-plane.

This model has to be coupled to appropriate carrier transport equations and a thermal model as presented in the next chapters 3 and 4. It describes well essential qualitative characteristics of real devices. A description for the retrieval of corresponding characteristics and distributions that are comparable to measurements is given.

Chapter 3

Carrier transport model

In diode lasers occupation inversion is achieved by applying a forward bias at the n- and p-contacts, so that the difference of the quasi Fermi energies for electrons and holes becomes greater than the energy difference of conduction and valence band edges. In high power diode lasers not only the spatial and temporal changes of carriers within the active region but also the transport of charged carriers from the contacts needs to be considered as processes such as current spreading and spatial hole burning depend on it and greatly influence the lasing operation.

This bipolar carrier transport can be described by the full drift-diffusion equations with sufficient accuracy, where drift is generated by the gradient of an electric potential and diffusion by the gradient of the carrier concentration, Eqs. (3.23) and (3.24). These equations are complemented by continuity equations connecting spatial changes of the current density with temporal changes of the carrier density and recombination, Eqs. (3.1) and (3.2), and the Poisson equation for the electrostatic potential (3.3).

Additionally under continuous-wave operation the transport of charge carriers and photons must be consistently formulated with the temperature flow, so that a self-consistent energy-transport model is necessary. Such a model has been given in [45] using Boltzmann statistics and equal electron, hole and lattice temperatures.

In the model presented here the active QW region is treated as a thin layer. The QW properties such as the Fermi-potential (3.12) in the quantum well as well as the coefficients describing optical gain (2.31) and carrier induced refractive index changes (2.29) are derived using a microscopic gain model [56]. Thus the influence of the Fermi-potential on the stimulated recombination and the resulting influence on the optical field is properly accounted for.

Laser simulation tools solving the full drift-diffusion equations [21, 19] are based on a stationary approximation and use for the description of the optical field either an expansion into linear waveguide modes [21] or a beam propagation method [19]. However, as a consequence of the dynamic behavior of broad-area (BA) lasers these approximations are questionable. Due to computer restrictions, until now simulation tools based on the traveling-wave model for the optical field do not solve the time-dependent drift-diffusion equations. They usually use a lateral diffusion equation for the excess carriers in the active layer with a spatially constant injection current density [28, 29, 30]. Such a constant-injection-current-density model, which is even used by stationary simulation tools [31, 32] oversimplifies the current flow and carrier transport in the device, because current spreading and current self-distribution are not taken into account [33].

In this chapter the basic drift-diffusion equations are presented as self-consistently formulated with the temperature flow in the energy-transport model given in [45, 46]. They will be reduced to an effective diffusion equation in the active region with a carrier density dependent diffusion coefficient. In contrast to assuming a constant injection current density, an advanced model will be gained from the drift-diffusion equations by neglecting recombination in the bulk layers and by assuming an infinitely high conductivity in the n-doped region. The resulting Laplace equation for the quasi-Fermi potential of the holes in the p-doped region can be solved either approximately as proposed in [57] (Joyce model) or exactly [18, 25]. A comparison of the different injection current density models will be given.

3.1 Basic drift-diffusion model

In a bulk semiconductor material the particle current flow is governed by the continuity equations for electrons and holes,

$$\nabla \cdot \vec{j_n} = e\,(R + \partial_t n)\,, \tag{3.1}$$

$$\nabla \cdot \vec{j_p} = -e\,(R + \partial_t p)\,. \tag{3.2}$$

Here $\vec{j_n}$ and $\vec{j_p}$ are the current densities for electrons and holes, respectively, and R the recombination rate. The electrostatic field itself is affected by the charge distribution of mobile (n and p) and fixed (n_A and p_D) carrier densities. The corresponding electrostatic potential φ solves the Poisson equation

$$-\vec{\nabla}(\varepsilon_0\varepsilon_s\vec{\nabla}\varphi) = e(p - n + p_D - n_A) \tag{3.3}$$

with the relative static dielectric constant ε_s. As a consequence of the Poisson equation (3.3) and the continuity equations (3.1) and (3.2) the current conservation equation holds,

$$\nabla \cdot \vec{j} = 0 \quad \text{with} \quad \vec{j} = -\varepsilon_0\varepsilon_s\nabla(\partial_t\varphi) + \vec{j}_n + \vec{j}_p. \tag{3.4}$$

In the model presented here in all layers charge neutrality

$$e(p - n + p_D - n_A) = 0 \tag{3.5}$$

is assumed, so that the electron and hole densities are given by

$$n = n_0 + N \quad \text{and} \quad p = p_0 + N, \tag{3.6}$$

respectively, with n_0, p_0 being the equilibrium densities and N the excess carrier density. At hetero-boundaries space charge layers develop, so that the assumption of charge neutrality is not valid in these regions. However, we are interested in laser operation far above threshold, where under forward bias the space charge layers become sufficiently thin so that Eq. (3.5) applies.

As already pointed out in Chapter 1 in diode lasers heterostructures with different band gap materials are used to restrict the carrier flow and to obtain a large carrier density in the active region. In Fig. 3.1 an exemplary energy band diagram and carrier density distribution of a super large optical cavity as employed in BA lasers is shown. By applying a forward bias U the quasi Fermi energies for electrons E_n^F and holes E_p^F

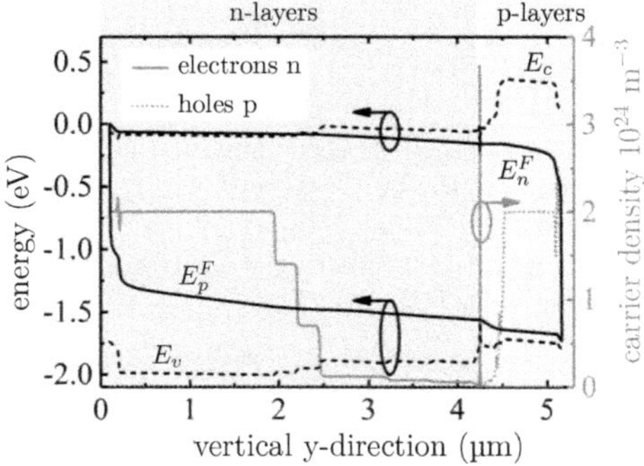

Figure 3.1: Exemplary energy band diagram (black left axis) and carrier density distribution (red right axis) as function of the vertical direction simulated for structure II (Tabs. A.3 and A.4) at an injection current density of $j = 5 \cdot 10^7$ A/m^2 using the laser simulation tool WIAS-TeSCA [58], which solves the full drift-diffusion equations in vertical direction assuming stationary laser operation.

are located in the conduction E_c and valence E_v bands, respectively, so that occupation inversion is gained.

For parabolic bands the Fermi integrals $\mathcal{F}_i$ with $i = 1/2$ for bulk and $i = 0$ for quantum well active regions connects the carrier densities n and p and the effective densities of states N_c and N_v with the exponents of the Fermi function $(e\varphi_n + e\phi - E_c)/k_BT$ and $(E_v + e\varphi_p - e\phi)/k_BT$, where $\varphi_n = -E_n^F/e$ and $\varphi_p = -E_p^F/e$ are the quasi Fermi potentials of electron and holes (E_n^F and E_p^F the corresponding Fermi energies) and E_c and E_v conduction and valence band edge energies,

$$\mathcal{F}_i(x) = y \tag{3.7}$$

with

$$x = \left(\frac{e\varphi - e\varphi_n - E_c}{k_BT}\right) \quad \text{or} \quad x = \left(\frac{E_v + e\varphi_p - e\varphi}{k_BT}\right) \tag{3.8}$$

and

$$y = \frac{n}{N_c} \quad \text{or} \quad y = \frac{p}{N_v} \tag{3.9}$$

respectively, with the conduction and valence band effective density of states N_c and N_v. For the bulk case ($i = 1/2$) and parabolic bands the effective density of states are given as

$$N_c = 2\left(\frac{2\pi m_n^* kT}{h^2}\right)^{3/2} \quad \text{and} \quad N_v = 2\left(\frac{2\pi m_p^* kT}{h^2}\right)^{3/2}, \tag{3.10}$$

where m_n^* and m_p^* are the effective masses of electrons and holes respectively.

According to Eq. (3.7) the corresponding inverse Fermi integrals $\mathcal{F}_i^{\text{inv}}$ yield the exponents of the Fermi function

$$x = \mathcal{F}_i^{\text{inv}}(y) \tag{3.11}$$

and thus the Fermi voltage is given by

$$\varphi_{\mathrm{F}} = \frac{k_{\mathrm{B}}T}{e}\left[\mathcal{F}_i^{\mathrm{inv}}\left(\frac{p}{N_{\mathrm{v}}}\right) + \mathcal{F}_i^{\mathrm{inv}}\left(\frac{n}{N_{\mathrm{c}}}\right)\right] + \frac{E_{\mathrm{g}}}{e} \tag{3.12}$$

where $E_{\mathrm{g}} = E_{\mathrm{c}} - E_{\mathrm{v}}$ is the energy gap. The Fermi integral can't be evaluated analytically, but has to be approximated. For the bulk case with $i = 1/2$ different approximations are discussed in Appendix C.

In the presence of a temperature gradient the electron and hole current densities generally have to be calculated by [45]

$$\vec{j}_n = -\,\sigma_n(\vec{\nabla}\varphi_n - P_n\vec{\nabla}T) \tag{3.13}$$

$$\vec{j}_p = -\,\sigma_p(\vec{\nabla}\varphi_p + P_p\vec{\nabla}T), \tag{3.14}$$

where σ_n and σ_p are the electrical conductivities and P_n and P_p are the Seebeck coefficients or thermoelectric powers being the entropies per particle divided by the elementary charge e. The Seebeck coefficients P_n and P_p in units of VK^{-1} can be derived as [45]

$$P_n = \frac{k_{\mathrm{B}}}{e}\left[1 + R_n - \frac{\partial_T E_{\mathrm{c}}}{k_{\mathrm{B}}} + \frac{e\varphi_n + E_{\mathrm{c}} - e\varphi}{k_{\mathrm{B}}T}\right], \tag{3.15}$$

$$P_p = \frac{k_{\mathrm{B}}}{e}\left[1 + R_p + \frac{\partial_T E_{\mathrm{v}}}{k_{\mathrm{B}}} - \frac{e\varphi_p + E_{\mathrm{v}} - e\varphi}{k_{\mathrm{B}}T}\right] \tag{3.16}$$

where n, p are kept constant in the differentiation. The functions R_n and R_p are given by the temperature derivatives of the inverse Fermi integrals,

$$R_n = -k_{\mathrm{B}}T\partial_T\mathcal{F}_i^{\mathrm{inv}}\left(\frac{n}{N_{\mathrm{c}}}\right) \tag{3.17}$$

$$R_p = -k_{\mathrm{B}}T\partial_T\mathcal{F}_i^{\mathrm{inv}}\left(\frac{p}{N_{\mathrm{v}}}\right). \tag{3.18}$$

For the bulk case ($i = 1/2$), Boltzmann statistics, parabolic bands and temperature-independent electron and holes masses R_n and R_p are equal to 3/2.

If the coefficients in front of the temperature derivatives in Eqs. (3.13) and (3.14) are derived from the Boltzmann equation in relaxation time approximation, then the factorization $\sigma_n \cdot P_n$ and $\sigma_p \cdot P_p$ is only possible for parabolic bands and Boltzmann statistics. The same holds for the factorization of the electrical conductivities into products of carrier-density independent mobilities and carrier densities, $\sigma_n = e\mu_n \cdot n$ and $\sigma_p = e\mu_p \cdot p$. Furthermore, the magnitudes of P_n, P_p similarly as μ_n, μ_p depend on the scattering processes involved [59]. For example, if the dependence of the relaxation time on the energy is given by $\tau_0[E/(k_{\mathrm{B}}T)]^r$ where r ranges typically between $-3/2$ and $+3/2$, then $R_n = R_p = 3/2 + r$ in (3.15) and (3.16). Thus $R_n = R_p = 3/2$ holds only for an energy-independent relaxation time. The temperature derivatives $\partial_T E_{\mathrm{c}}$ and $\partial_T E_{\mathrm{v}}$ of the conduction and valence band edges are often not included in the definition of the Seebeck coefficients.

The recombination rate R in Eqs. (3.1) and (3.2) consists of a radiative R_{rad} and a non-radiative part $R_{\text{non-rad}}$ (including Shockley-Read-Hall R_{SRH} and Auger recombination R_{Aug}). The radiative part can again be divided onto a spontaneous R_{sp} and stimulated recombination rate R_{st}, so that

$$R(t,\vec{r}) = R_{\text{non-rad}}(t,\vec{r}) + R_{\mathrm{sp}}(t,\vec{r}) + R_{\mathrm{st}}(t,\vec{r}). \tag{3.19}$$

Shockley-Read-Hall (SRH) recombination involves trap levels inside the band gap that are generated by crystal defects. Electrons from the conduction band and holes from the valence band can be captured by those deep levels thereby transferring their energy to the lattice. SRH recombination is dominated by the minority carriers and depends on capture coefficients, trap densities and energies, but is often described by the SRH lifetime τ_{SRH},

$$R_{\mathrm{SRH}} = \frac{np - n_0 p_0}{\tau_{n,\mathrm{SRH}}(p + p_1) + \tau_{p,\mathrm{SRH}}(n + n_1)} \tag{3.20}$$

with the equilibrium electron n_0 and hole p_0 density and $n_1 = N_c \exp((E_t - E_c)/k_B T)$ and $p_1 = N_v \exp((E_v - E_t)/k_B T)$ with the trap energy E_t. τ_{SRH} depends on the type of defects and the fabrication process and is difficult to determine. Thus it should be fit to experimental data [60].

The rate of radiative spontaneous recombination is often characterized by a material coefficient B and written as

$$R_{\mathrm{sp}} = B(np - n_0 p_0) \tag{3.21}$$

with the equilibrium electron and hole densities n_0 and p_0, respectively.

Auger recombination transfers the excess energy and momentum of the recombination of an electron-hole pair to another electron within the valence or conduction band and may involve different valence bands and the interaction with phonons. It is given by

$$R_{\mathrm{Aug}} = (C_n n + C_p p)(np - n_0 p_0) \tag{3.22}$$

with the Auger coefficients C_n and C_p for electrons and holes, respectively. The coefficients B and $C_{n,p}$ for spontaneous and Auger recombination are often determined as the second- (BN^2), respective third- ($C_{n,p}N^3$) order coefficients in a polynomial fit to measurements as a function of average carrier concentration N. Reported values at room-temperature can be found in e.g. [60].

The rate of stimulated recombination is the rate by which the energy density of the optical field changes by stimulated emission or absorption of a photon due to transitions between the conduction and valence bands, divided by the energy of the emitted or absorbed photon and follows from Eqs. (2.14) and (2.16) to Eq. (2.18).

3.2 Reduction to effective diffusion equation and models for the injection current

We will now proceed with the bulk case ($i = 1/2$). As discussed above the active QW region is treated as a thin layer. The QW properties are however still incorporated in the model by a fit of model parameters to a microscopic gain model [56].

Furthermore vanishing Seebeck coefficients are assumed. In this way the equations for the electronic and heat model are also valid in the non-Boltzmann case. As a consequence, a feedback of temperature on the current (Seebeck effect), is neglected. However, as current conservation and charge neutrality must hold, this only means slightly altered voltage gradients at hetero-boundaries and has no influence on the laser operation. The consequences on the heat generation will be discussed in Section 4.2.2.

Using Eqs. (3.7)-(3.9) the electron (3.13) and hole (3.14) current flow can be separated into parts driven by diffusion and drift [61],

$$\vec{j_n} = n \cdot \mu_n \nabla (E_c - e\varphi) + eD_n \vec{\nabla} n, \tag{3.23}$$

$$\vec{j_p} = p \cdot \mu_p \nabla (E_v - e\varphi) - eD_p \vec{\nabla} p, \tag{3.24}$$

with the concentration dependent diffusion coefficient for electrons D_n and holes D_p

$$D_n = \mu_n \frac{k_B T}{e} \frac{n}{N_c} \mathcal{F}_i^{\mathrm{inv}\prime} \left(\frac{n}{N_c} \right), \tag{3.25}$$

$$D_p = \mu_p \frac{k_B T}{e} \frac{p}{N_v} \mathcal{F}_i^{\mathrm{inv}\prime} \left(\frac{p}{N_v} \right) \tag{3.26}$$

where $\mathcal{F}_i^{\mathrm{inv}\prime}$ is the derivative of the inverse Fermi integral with respect to n or p.

As the device is longitudinally much further extended than laterally, longitudinal carrier transport plays a minor role and the study will be restricted to vertical-lateral carrier transport only.

One of the basic assumptions for the presented carrier transport model is that the electron conductivity in the n-doped region is much larger than the hole conductivity in the p-doped region. Hence, by assuming $\sigma_n \to \infty$ it follows $\nabla \varphi_n \to 0$ so that

$$\varphi_n \approx \text{const.} = 0 \tag{3.27}$$

can be set. As a consequence the electron current $\vec{j_n}$ is composed only of a vertical component $j_{n,y}$, that remains constant in vertical direction (see Fig. 3.3).

Furthermore recombination in the bulk layers is neglected, which results in vanishing vertical leakage currents. All vertical structures discussed in this thesis are appropriately designed to reduce vertical leakage currents due to high potential barriers and thin p-layers, so that this approximation is valid.

3.2.1 Carrier transport in the active region

The active region is undoped, thus within the active region, $p_D = n_A = 0$. The carrier density is given by the mobile charges (3.6). Following [62] an equation for the lateral carrier transport within the active region is derived, which is considered to be thin enough so that vertical carrier transport can be averaged over.

Considering (3.27) and inserting (3.25) into (3.23) we obtain

$$\partial_x \left(\frac{E_c}{e} - \varphi \right) = -\partial_x \left(\frac{k_B T}{e} F_{1/2}^{\mathrm{inv}} \left(\frac{n}{N_c} \right) \right). \tag{3.28}$$

We exploit the parallelism of band edges ($\nabla E_c = \nabla E_v$) and insert (3.28) into (3.24) to obtain an equation for the lateral current density,

$$j_{p_x} = -\sigma_p \partial_x \left(\frac{k_B T}{e} F_{1/2}^{\mathrm{inv}} \left(\frac{n}{N_c} \right) \right) - \sigma_p \partial_x \left(\frac{k_B T}{e} F_{1/2}^{\mathrm{inv}} \left(\frac{p}{N_v} \right) \right). \tag{3.29}$$

With the definition of the Fermi integral (3.7) we obtain,

$$j_{p_x} = -\sigma_p \partial_x \left(-\varphi_n - E_c/e + \varphi + E_v/e - \varphi + \varphi_p \right) \tag{3.30}$$

$$= -\sigma_p \partial_x (\varphi_p - \varphi_n) = -\sigma_p \partial_x \varphi_F \tag{3.31}$$

$$= -eD_{\mathrm{eff}} \partial_x N \tag{3.32}$$

with an effective diffusion coefficient

$$D_{\text{eff}} = \sigma_p \partial_N \varphi_F / e \tag{3.33}$$

with

$$\partial_N \varphi_F = \frac{k_B T}{e} \left[\frac{1}{N_c} F_{1/2}^{-1'} \left(\frac{N(r,t)}{N_c} \right) + \frac{1}{N_v} F_{1/2}^{-1'} \left(\frac{N(r,t)}{N_v} \right) \right], \tag{3.34}$$

and conductivity

$$\sigma_p = e\mu_p (p_0 + N). \tag{3.35}$$

Often a carrier independent constant diffusion coefficient is considered [32, 27, 28] that is derived in the Boltzmann approximation as

$$D_c = 2\mu_p k_B T / e. \tag{3.36}$$

In Fig. 3.2 the effective diffusion coefficient calculated in the Joyce-Dixon approximation [63] (*cf.* Appendix C) is displayed with the diffusion constant D_c for comparison. Already at moderate carrier densities of $N = 3 \cdot 10^{24}$ m^{-3}, D_{eff} is 50% higher than D_c, rising to $5D_c$ at extremely high densities of $N = 10 \cdot 10^{24}$ m^{-3}.

Inserting (3.32) into the continuity equation (3.2) gives

$$\partial_t N = \partial_x (D_{\text{eff}} \partial_x N) - \frac{\partial_y j_{p_y}}{e} - R. \tag{3.37}$$

Due to the thin active layer the divergence of the carrier density in vertical y-direction is assumed to be of marginal importance. Thus the process of carrier transport can be modeled by a mean value for the carrier density (sheet density divided by the QW thickness). Averaging (3.37) across the active region with thickness d and neglecting the vertical hole leakage current, which is well valid due to low hole mobility and usually high potential barriers, yields,

$$\begin{aligned} \partial_t N &= \partial_x (D_{\text{eff}} \partial_x N) - \frac{1}{ed} \int_0^d \partial_y j_{p_y} dy - R \\ &= \partial_x (D_{\text{eff}} \partial_x N) + \frac{1}{ed} \Big(\underbrace{-j_{p_y}(d)}_{=j} - \underbrace{(-j_{p_y}(0))}_{=j_p^{\text{leak}} \to 0} \Big) - R \qquad (3.38) \\ &= \partial_x (D_{\text{eff}} \partial_x N) + \frac{j}{ed} - R. \qquad (3.39) \end{aligned}$$

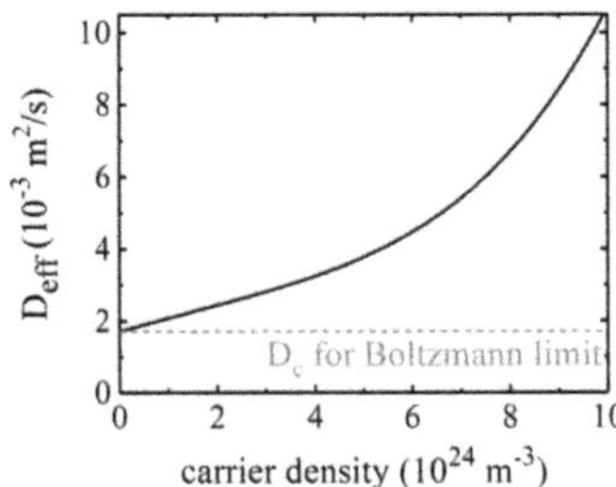

Figure 3.2: Effective diffusion coefficient D_{eff} as function of carrier density and comparison with diffusion constant in Boltzmann limit D_c.

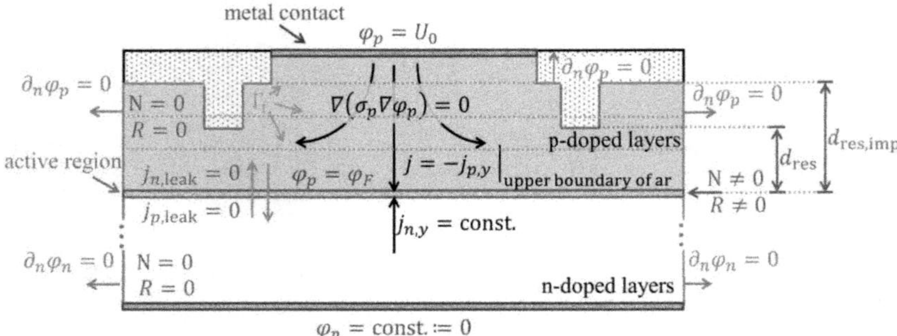

Figure 3.3: Sketch of the lateral-vertical (x, y)-cross section of the laser. The domain for which the Laplace equation (3.43) is solved is marked grey. Black dotted regions are non conducting.

Note that the injection current density j is given by the negative hole current density at the upper boundary of the active region (see Fig. 3.3),

$$j = -j_{p_y}|_{\text{upper boundary of ar}} \tag{3.40}$$

and can be calculated by different models as described in the following section. Inserting (3.6) into (3.19) and assuming $N \gg n_0, p_0$, appropriate for undoped active regions, the recombination rate becomes

$$R = AN + BN^2 + CN^3 + R_{\text{st}}, \tag{3.41}$$

where $A = 1/(\tau_{n,\text{SRH}} + \tau_{p,\text{SRH}})$, B, $C = C_n + C_p$ and R_{st} are the the Shockley-Read-Hall, the spontaneous radiative, the Auger and the stimulative recombination rate from (2.39), respectively.

3.2.2 Models for the injection current density

Due to assumption of an infinite electron conductivity in the n-doped region, only the current flow in the p-doped region (grey in Fig. 3.3) needs to be considered. The lower boundary is given by the upper boundary of the active region (striped red in Fig. 3.3). The upper boundary is defined by the metal contact in the center. The other boundaries are given by adjacent regions with a vanishing conductivity (black dotted in Fig. 3.3). Such regions for current confinement can be created by e.g. implantation, diffusion or impurity induced disordering. The distance between the upper boundary of the active region and the lower boundary of the implanted region is denoted by the residual layer thickness $d_{\text{res,imp}}$. If index guiding trenches are etched additionally, those regions are also non-conducting and the distance between the upper boundary of the active region and the lower boundary of the index guiding trenches is denoted as d_{res}.

By neglecting recombination in the bulk layers, the drift-diffusion equation in the p-doped layer (3.2) results in

$$\nabla \vec{j}_p = 0. \tag{3.42}$$

The neglect of recombination in the bulk layers follows from vanishing vertical leakage currents. This is well valid for appropriately designed vertical structures with high potential barriers and thin p-layers.

By inserting (3.14) with $P_p = 0$ we obtain the Laplace equation for the hole Fermi potential,

$$\nabla(\sigma_p \nabla \varphi_p) = 0. \tag{3.43}$$

At the upper metal contact the potential is given by the applied voltage $\varphi_p = U$, at the lower boundary to the active region the potential is given by the Fermi potential $\varphi_p = \varphi_F$ (3.12). At all other device boundaries the derivative of the hole quasi Fermi potential normal to the device surface vanishes, $\partial_n \varphi_p = 0$ (see Fig. 3.3). At internal vertical hetero-interfaces Γ_i in Fig. 3.3 the hole quasi Fermi potential and current density are continuous, i.e. $\varphi_p|_{y\in\Gamma_i^+} = \varphi_p|_{y\in\Gamma_i^-}$ and $\sigma_p \partial_n \varphi_p|_{y\in\Gamma_i^+} = \sigma_p \partial_n \varphi_p|_{y\in\Gamma_i^-}$, respectively. In the Laplace model the injection current density entering Eq. (3.39) is accordingly given by

$$j(x) = \sigma_p \partial_y \varphi_p|_{y=y_{\mathrm{ar}}} \tag{3.44}$$

where y_{ar} gives the y-position at the upper boundary of active region. The Joyce model [57] is based on an approximate solution of the Laplace equation for the quasi-Fermi potential of the holes in the p-doped region and describes the injection current density as

$$j(x) = \begin{cases} (U - \varphi_F)/r_{\mathrm{s}} & \text{below the contact stripe} \\ \partial_x^2 \varphi_F / \Omega & \text{beside the contact stripe,} \end{cases} \tag{3.45}$$

where

$$r_{\mathrm{s}} = \int_{\text{p-layers}} \frac{1}{\sigma_p(y)} dy \tag{3.46}$$

is the series resistance related to the contact area and

$$\Omega = \frac{1}{\int_{\text{p-layers}} \sigma_p(y) dy} \tag{3.47}$$

is the sheet resistance.

The model of a spatially constant current injection density j_0 used by many simulation tools is given by

$$j(x) = \begin{cases} j_0 & \text{below the contact stripe} \\ 0 & \text{beside the contact stripe.} \end{cases} \tag{3.48}$$

To investigate the impact of the different injection current density models, the simulation results for a model structure are displayed in Fig. 3.4. The laser is optically guided by an index-guiding trench at $|x| \in [45:50]$ µm ($\Delta n_0 = -10^{-3}$) to avoid high amounts of optical pumping and carrier generation beside the stripe. However, the structure (structure II Tab. A.3) with injection stripe width $W = 90$ µm and length $L = 4$ mm is electrically only shallowly implanted ($d_{\mathrm{res,imp}} = 910$ nm) to observe the effect of the different injection current density models. The data was obtained by transient simulations with time period $\tau_{\mathrm{sim}} = 8$ ns. The data was averaged over the last $\tau_{\mathrm{av}} = 5$ ns of that time span. The near and far intensities are derived according to Eqs. (2.41) and (2.43), respectively.

The Joyce model has a discontinuity at the stripe edges and thus the calculated injection current density for the Joyce model yields unphysical high side peaks at the stripe edges, Fig. 3.4(a). Furthermore the spreading of the current into areas beside the injection stripe is underestimated. As a result, the carrier density distribution beside

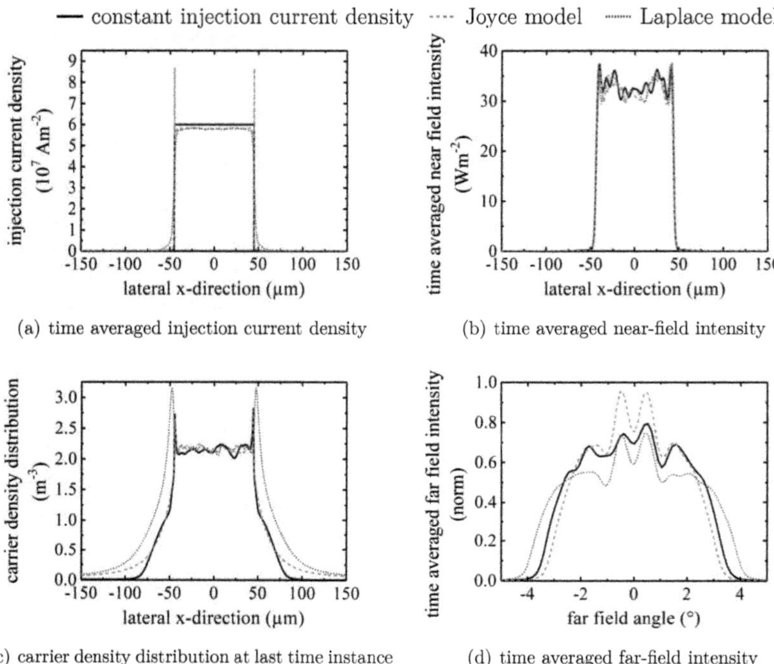

(c) carrier density distribution at last time instance

(d) time averaged far-field intensity

Figure 3.4: Comparison of different injection current density models for a total injection current of $I = 20$ A. Distributions are displayed for the front facet. Simulation time: $\tau_{\text{sim}} = 8$ ns. Time for averaging data: $\tau_{\text{av}} = 5$ ns.

the stripe is much higher for the Laplace model compared to the other models, Fig. 3.4(c).

Furthermore it is noteworthy that the carrier density distribution below the stripe is also affected by the choice of the injection current density model. The undulation of the carrier density below the stripe results from depletion of carriers due to stimulated emission and is termed spatial hole burning. For high values of the series resistivity r_s, spatial hole burning becomes more pronounced, because the injection current density approaches a constant value (Eq. (3.48) is a limit case of Eq. (3.45) for $r_s \to \infty$ and $\Omega \to \infty$). This can be seen by expanding φ_F in (3.45) around the average carrier density $\bar{N}$

$$j(N) = \bar{j} - \frac{\partial_N \varphi_F}{r_s}(N - \bar{N}). \tag{3.49}$$

For injection current density models that take into account current self-distribution, such as the Joyce (3.45) and Laplace model (3.43), a decreased carrier density leads to a drop of the Fermi-potential and a consequent increase of the injection current density at that position so that spatial hole burning is counteracted. As a result the constant injection current density model overestimates spatial hole burning, which is also visible in Fig. 3.4(c). This effect of current self-distribution is also discussed in e.g. [33, 64].

Although the time averaged near-field intensity which is stabilized by the index-guiding trenches remains largely unaffected by the choice of injection current density model, Fig. 3.4(b), the far-field intensity depends on it, Fig. 3.4(d). This effect can easily be understood in the mode picture by decomposing the time dependent field intensity into hypothetical steady state modes as will be discussed in more detail in Section 6.3.3. Here, it shall only be anticipated that the contribution of higher order lateral modes to the lasing process is higher for the constant injection current density model than for the Joyce model and even higher for the Laplace model, which is why their far field is broader.

Excitation of higher order lateral modes is aided by lateral spatial hole burning and high spreading currents, see Section 6.3.3. Thus a proper description of spatial hole burning as well as current spreading is important for the investigation of phenomena in BA lasers and design concepts for the improvement of their beam quality. Whereas both the Joyce and Laplace model describe the effect of current self-distribution, only the Laplace model properly accounts for current spreading into regions beside the injection stripe.

3.3 Summary

In this chapter the basic drift-diffusion equations are revisited as presented in [45]. They are reduced to an effective diffusion equation in the active region with a carrier density dependent diffusion coefficient. The correct description of the injection current density entering this effective diffusion equation is important to consider spatial hole burning by current-self-distribution and the lateral spreading of the current into areas beside the injection stripe.

If only a constant injection current density is assumed below the injection stripe, current spreading is neglected and spatial hole burning overestimated. An improved model for the injection current density can be gained by neglecting recombination in the bulk layers and by assuming an infinitely high conductivity in the n-doped region. The resulting Laplace equation for the quasi-Fermi potential of the holes in the p-doped region can be solved either approximately as proposed in the Joyce model [57] or exactly [18, 25].

One further assumption of this model bases on the neglect of vertical leakage currents. All vertical structures discussed in this thesis are appropriately designed to reduce vertical leakage currents due to high potential barriers and thin p-layers, so that this approximation is valid.

In the Joyce model current self-distribution is correctly described and the strength of spatial hole burning depends on the series resistivity r_s. However, the Joyce model has a discontinuity at the edges of the injection stripe and thus generates unphysical high side peaks of the injection current density. Furthermore current spreading is underestimated. Thus only the Laplace model properly describes the injection current density entering the lateral diffusion equation and will be used in the following to simulate laser operation.

The derived carrier transport equations and the equations for the optical field as discussed in Chapter 2 have to be self-consistently coupled to the heat transport as discussed in the next chapter.

Chapter 4

Heat model

For the accurate description of continuous-wave (CW) laser operation the transport of the charged carriers (electrons and holes) and the photons must be consistently formulated with the temperature flow. Only in this way, self-heating effects such as thermal roll-over and thermal waveguiding can be described properly. Under pulsed operation self-heating is usually neglected because thermal build-up times are much longer than the pulse lengths. However, the heat is generated near to the active layer in the same region where the guided wave is localized. This region is small and its thermal build-up time is much shorter than that of the whole device. Accordingly under pulsed operation short-time local heating can influence the optical pulse formation although time-averaged heating is negligible. Results supporting this claim, have been published in [4] and are elaborated in more detail in Section 6.4.1.

Due to the fluctuation of field intensity and carrier density the heat sources and the resulting temperature profile are non-stationary [4]. Theoretically, a time-dependent temperature has been considered in early models as presented e.g. in [34, 35]. They concentrated on a sophisticated microscopic description of the processes in the active layer. However, besides requiring enormous computer resources even for nanosecond transients, outer parts of devices are disregarded although a considerable portion of the heat is generated here. Heat flow was replaced by a simple local relaxation of temperature towards an ambient temperature. These features prevent an application in device design.

A derivation of a self-consistent energy-transport model by applying fundamental thermodynamic principles has been given in [46]. A similar model is derived in [45] using Boltzmann statistics. In these models electron, hole and lattice temperatures are assumed to be equal, opposed to other energy-transport models [65, 66]. A more general energy-transport model for semiconductor devices has been derived in [67] taking into account Fermi statistics and different temperatures for the charged carriers and the lattice, but disregarding optical fields.

The thermal model presented here bases on [45, 46] and the energy-transport model is summarized in Section 4.1 paying particular attention to a consistent formulation with the model for the optical field given in Chapter 2. In the following Section 4.2 approximations are made to the energy-transport model. Particularly vanishing Seebeck coefficients are assumed and the heat sources are formulated in consistency with the optical longitudinal-lateral approximate equations and the carrier transport model. Their vertical and longitudinal-lateral distribution as well as the relative share of the different heat sources to the total heating is discussed. Energy conservation for the employed equations is derived in Section 4.3. For the treatment of pulsed operation in Section 4.4 inherent traits of heat generation and conduction can be exploited so that the temperature induced refractive index can be self-consistently derived from the heat sources for short pulse length. CW operation is solved by an iterative coupling of the electro-optical models to the heat transport equations as described in Section 4.5 resulting in a time-dependent quasi-three-dimensional opto-electronic and thermal model. The presented results are already partly published in Refs. [4] and [49].

4.1 Basic equations

In this section the basic heat-flow equations as given in [45, 46] and the physical meaning of the different heat source density contributions are discussed.

The heat flow is determined by the heat-flow equation [45, 46]

$$c_h \partial_t T - \vec{\nabla}(\kappa_L \vec{\nabla} T) = h(N, \vec{j}, \|u\|^2) \tag{4.1}$$

with the heat capacity c_h, lattice thermal conductivity κ_L and heat source density h.

The boundary conditions for (4.1) are

$$\begin{aligned} \kappa_L \partial_n T &= -(T - T_{HS})/r_{th} \quad \text{for } (x,y,z) \in \text{heat sink} \\ \partial_n T &= 0 \qquad\qquad\qquad \text{for } (x,y,z) \in \text{other outer bounds} \end{aligned} \tag{4.2}$$

where r_{th} is the inverse heat transfer coefficient. Here and in what follows, T is the absolute temperature and T_{HS} the temperature of the heat sink, which also serves as reference for parameter values.

The heat source density can be witten as [46]

$$\begin{aligned} h =& \frac{1}{\sigma_n}\vec{j}_n^2 + \frac{1}{\sigma_n}\vec{j}_n^2 + T(\vec{j}_n \vec{\nabla} P_n - \vec{j}_p \vec{\nabla} P_p) \\ & - T(\partial_T \varphi_n|_{n,p} - P_n) \cdot \vec{\nabla} \vec{j}_n - T(\partial_T \varphi_p|_{n,p} + P_p) \vec{\nabla} \vec{j}_p \\ & + e(T\partial_T \varphi_n|_{n,p} - \varphi_n - T\partial_T \varphi_p|_{n,p} + \varphi_p) R - \langle \vec{\nabla} \cdot \vec{S}_{\text{rad}} \rangle_t - \langle \partial_t u_{\text{rad}} \rangle_t. \end{aligned} \tag{4.3}$$

Here, no assumption regarding the Seebeck coefficients P_n and P_p was made. Note, that different expressions for the Seebeck coefficients exist [68]: A theoretical Seebeck coefficient can for example be derived according to the Kelvin formula [68] from the quasi-Fermi potentials as $P_n = \partial_T \varphi_n|_{n,p}$ and $P_p = -\partial_T \varphi_p|_{n,p}$. However, these expressions miss many body effects due to different scattering mechanisms. An alternative way of deriving the Seebeck coefficients is from the Boltzmann equation in relaxation time approximation [69] assuming parabolic bands and Boltzmann statistics as discussed in Section 3.1. Considering these effects a more complete relation is given by [69]

$$P_n = \partial_T \varphi_n|_{n,p} + k_B/e(1+r) \tag{4.4}$$

and

$$P_p = -\partial_T \varphi_p|_{n,p} + k_B/e(1+r), \tag{4.5}$$

where r is the exponent of the energy dependence of the relaxation time. Thus the commonly used expressions for the Seebeck coefficients (3.15) and (3.16) are obtained. Using Eqs. (4.4) and (4.5) the heat source density is given by

$$\begin{aligned} h =& \frac{1}{\sigma_n}\vec{j}_n^2 + \frac{1}{\sigma_n}\vec{j}_n^2 + T(\vec{j}_n \vec{\nabla} P_n - \vec{j}_p \vec{\nabla} P_p) \\ & + k_B T/e(1+r) \cdot \vec{\nabla}(\vec{j}_p - \vec{j}_n) - 2k_B T(1+r) R \\ & + e(TP_n - \varphi_n + TP_p + \varphi_p) R - \langle \vec{\nabla} \cdot \vec{S}_{\text{rad}} \rangle_t - \langle \partial_t u_{\text{rad}} \rangle_t. \end{aligned} \tag{4.6}$$

All quantities appearing in Eq. (4.6) have been introduced in the preceding chapters, particularly a balance equation for the radiative energy density $\vec{\nabla} \cdot \vec{S}_{\text{rad}} + \partial_t u_{\text{rad}}$ is given by Eq. (2.16). In the following the heat source density terms will be discussed separately.

The first and second term in Eq. (4.6) correspond to Joule heat

$$h_{\text{J}} = \frac{1}{\sigma_n}\vec{j}_n^2 + \frac{1}{\sigma_p}\vec{j}_p^2 \tag{4.7}$$

generated by scattering of the carriers on phonons resulting in a energy loss to the lattice.

The third term describes Thomson-Peltier heat

$$h_{\text{TP}} = T(\vec{j}_n \vec{\nabla} P_n - \vec{j}_p \vec{\nabla} P_p) \tag{4.8}$$

which is generated by a current flow along the gradients of the Seebeck coefficients.

The second line in (4.6) only gives a contribution for the non-stationary case and is often neglected in device simulation, as it is generally small. It is attributed to thermo-diffusion and similarly to the Seebeck coefficients (4.4) and (4.5) when derived from the Boltzmann equation in relaxation time approximation, many-body effects due to different scattering mechanisms are taken into account. Considering (3.1) and (3.2) it can be rewritten as,

$$(1+r) \cdot [k_{\text{B}}T/e\vec{\nabla} \cdot (\vec{j}_n - \vec{j}_p) - 2k_{\text{B}}TR] = k_{\text{B}}T(1+r)(\partial_t n + \partial_t p). \tag{4.9}$$

The last term is due to contributions of the recombination of electron-hole pairs which sets free the total electron $TP_n - \varphi_n$ and hole $TP_p + \varphi_p$ energy, that is either transferred to the lattice as heat or transferred to the radiative field. The latter part is described by the term $\langle \vec{\nabla} \cdot \vec{S}_{\text{rad}} \rangle_t + \langle \partial_t u_{\text{rad}} \rangle_t$, which has to be subtracted from the source term for energy conservation. Here, $\langle \rangle_t$ denotes the temporal average over several oscillation cycles. Thus the last line not only includes recombination heat, but also heat generated by the reabsorption of radiation.

Generally the radiative energy density u_{rad} and flux $\vec{S}_{\text{rad}}$ have contributions from spontaneous and stimulated emission. The term $\varepsilon_0 c\bar{n} \langle \text{Re}(E^{+*}F_{\text{sp}}^{+} + E^{-*}F_{\text{sp}}^{-}) \rangle_t$ in Eq. (2.16) accounts for the spontaneous emission into the lasing modes. Far above threshold this contribution is small and can be neglected. The spontaneous emission into all other modes is challenging to calculate. For simplification it is assumed that all spontaneous radiation is instantaneously reabsorbed in the cavity. A more sophisticated model is, however, not necessary as the error made due to this approximation is small as discussed in Section 4.2.1.

By inserting $R = R_{\text{non-rad}} + R_{\text{sp}} + R_{\text{st}}$ into the last line of Eq. (4.6) the different heating contributions can be distinguished. The recombination heat is given by

$$h_{\text{rec}} = e(TP_n + TP_p + \varphi_p - \varphi_n)(R_{\text{non-rad}} + R_{\text{sp}}) \tag{4.10}$$

and accounts for non-radiative recombination heat and heat due to the instantaneous reabsorption of the spontaneous radiation in the cavity. Furthermore by inserting Eq. (2.16) for $\vec{\nabla} \cdot \vec{S}_{\text{rad}} + \partial_t u_{\text{rad}}$ the defect heat

$$h_{\text{defect}} = [e(TP_n + TP_p + \varphi_p - \varphi_n) - \hbar\omega_0]R_{\text{st}} \tag{4.11}$$

and absorption heat

$$h_{\text{abs}} = \frac{\varepsilon_0 c}{2} n_{\text{r}} \alpha \|E\|^2 + \left(\frac{\varepsilon_0 c}{2}\right)^2 \bar{n} n_{\text{r}} \beta (\|E\|^4 + 4|E^+|^2|E^-|^2) \tag{4.12}$$

can be distinguished, respectively. The term (4.11) is caused by a possible incomplete energy transfer from the carrier ensemble to the radiation field during stimulated emission and is also referred to as quantum defect energy. Eq. (4.12) describes heat due to absorption of stimulated radiation.

Inserting the different contributions (4.7)-(4.12) into (4.6) gives the total heat source density as

$$h = k_{\text{B}} T(1 + r)(\partial_t n + \partial_t p) + h_{\text{J}} + h_{\text{rec}} + h_{\text{abs}} + h_{\text{defect}} + h_{\text{TP}}. \tag{4.13}$$

4.2 Approximate equations for the heat source density

To derive the heat equations in consistency with the optical and electronic model equations some approximations are made which will be discussed in the following.

4.2.1 Treatment of spontaneous emission

The spontaneous radiation could be treated similarly as the stimulated radiation but approximations have to be employed, because the field generated by spontaneous emission is more challenging to calculate. For simplification it is assumed that all spontaneous radiation is instantaneously reabsorbed in the cavity as recombination heat.

This approximation leads to an overestimation of the generated heat in the active region because spontaneously emitted photons are reabsorbed anywhere in the device or leave it. However, a detailed model is unnecessary because far above threshold where high power lasers are operated this heat contributes only to a small amount. For instance in the cases shown in Fig. 4.2 where laser operation with injection currents of 150 A and 20 A are discussed, this portion amounts to only 0.5% respective 6% of the total heat source density. Furthermore the vertical distribution of the heat sources in the device has only a small impact on the lateral temperature profile as was shown in [70].

4.2.2 Impact of vanishing thermoelectric effects on the heat generation

In the preceding chapter vanishing Seebeck coefficients are assumed to avoid major alterations of the electric model and to obtain validity in the non-Boltzmann case. The simplification of the electronic model has no influence on the laser operation as discussed in Chapter 3. However to maintain consistency, also in the heat model thermoelectric effects have to vanish, which means that Thomson-Peltier heat (4.8), heating due to thermo-diffusion (4.9) and the Seebeck coefficients in the recombination (4.10) and defect heat (4.11) term are neglected. The error made in the heat generation is small and estimated for an energy-independent relaxation time, i.e. $r = 0$, in the following.

The Thomson-Peltier heat (4.8) is generated by a current flow along the gradients of the Seebeck coefficients ∇P_n and ∇P_p, which give the increase or decrease of carrier potential with temperature, Eqs. (3.15) and (3.16). It can be separated into contributions due to Peltier heat and Thomson heat by applying the chain rule to the gradients,

$$h_{\text{Peltier}} = T\left(\vec{j}_n \partial_n P_n|_T \vec{\nabla} n - \vec{j}_p \partial_p P_p|_T \vec{\nabla} p\right) \tag{4.14}$$

$$h_{\text{Thomson}} = T\vec{\nabla} T\left(\vec{j}_n \partial_T P_n|_{(n,p)} - \vec{j}_p \partial_T P_p|_{(n,p)}\right). \tag{4.15}$$

The Peltier heat gives the more pronounced contribution and results from changes in carrier concentration for example at the interfaces of two materials with different doping concentrations or band edge energies. Thomson heat refers to changes of the carrier potential as a consequence of temperature gradients. In both cases energy is exchanged between carriers and lattice while a charge carrying current is overcoming the potential gradients.

Also in the recombination heat (4.10) a term accounts for the presence of Peltier coefficients,

$$h_{\text{rec},P_{n,p}} = e(TP_n + TP_p)R = (5k_BT + E_g - e\varphi_F)R, \tag{4.16}$$

with P_n and P_p from (3.15) and (3.16) for an energy-independent relaxation time, i.e. $R_n = R_p = 3/2$, and by neglecting the temperature change $\partial_T E_\text{c}$ and $\partial_T E_\text{v}$ of the conduction and valence band edges. Here eTP_n and eTP_p denote the excess energy of electrons and holes above the quasi-Fermi energies, respectively, and $R = R_\text{non-rad} + R_\text{sp} + R_\text{st}$ the total recombination rate.

The error made by neglecting (4.16) compared to the full model (4.10) can be easily assessed with

$$\sigma_{\text{h,rec}} = 1 - \frac{e\varphi_F}{5k_BT + E_g}. \tag{4.17}$$

For the structures investigated in this thesis with $E_g \approx 1.36$ eV, $N_c \approx 10^{24}$ m^{-3}, $N_v \approx 5 \cdot 10^{24}$ m^{-3} and φ_F from (3.12) in the Joyce-Dixon approximation (C.10), the deviation is in the range of 5% for an excess carrier density of $N = 3 \cdot 10^{24}$ m^{-3} and decreases with increasing current density because of an increased excess carrier density and a resulting rise of the quasi Fermi-potential φ_F. This error is small and thus the part (4.16) of the recombination is negligible.

In the presented model the term $2k_\text{B}T\partial_t N$ (4.9) attributed to thermo-diffusion only generates heat in the active region, where the excess carrier density is $N > 0$. It can be positive or negative depending on $\partial_t N$. For very short time instances of some picoseconds it can reach high values, especially during turn on. For example for the same structure as discussed in Section 4.2.3 at pulsed operation with 150 A injection current a maximum heat source density in the range of $3 \cdot 10^{15}$ Wm^{-3} is reached at the rear facet during turn on, which is comparable to other heat sources, Fig. 4.2(b). However, due to different signs in the contributions on average this heat source density remains well below $\langle 2k_\text{B}T\partial_t N\rangle_t < 10^{13}$ Wm^{-3} throughout the resonator and is thus negligible.

By assuming the Boltzmann case as well as temperature independence of conduction and valence band edges E_c and E_v the Thomson term can be simplified to $h_\text{Thomson} = 3k_\text{B}\vec{\nabla} T\left(\vec{j}_n - \vec{j}_p\right)/2e$. The main contribution to this heat term is generated in vertical

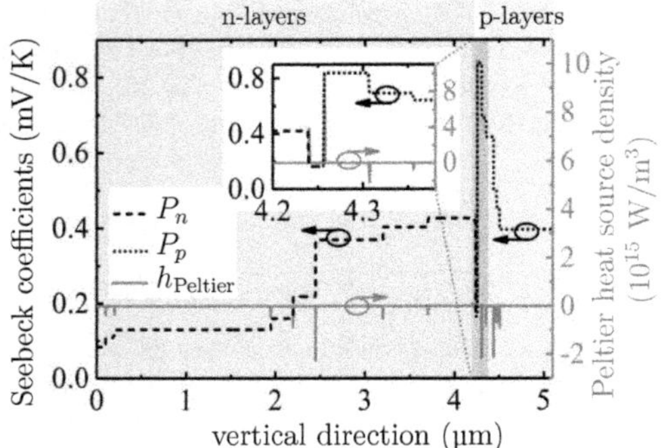

Figure 4.1: Black left axis: Vertical distribution of the Seebeck coefficients P_n and P_p calculated in the Boltzmann limit. Red right axis: Peltier heat source density generated at every hetero-interface due to a change of the Seebeck coefficients calculated for an injection current density of $j = 5 \cdot 10^7$ A/m^2.

direction, where the temperature gradient as well as current flow are highest. Depending on the position of the heat sink (device mounting) this contribution can be positive in the n-layers and negative in the p-layers or vice versa. Assuming current densities of $j_y < 10^8$ Am^{-2} and a vertical temperature gradient of $\partial_y T < 10^6$ Km^{-1} [71] this heat source density is in the range of $|h_{\text{Thomson}}| < 10^{11}$ Wm^{-3} and thus much smaller than other heat sources, see Fig. 4.2.

The Peltier heat (4.14) has contributions at vertical hetero-interfaces and results from doping gradients or different semiconductor material compositions, whereas its contribution in the (x, z)-plane can be neglected. By assuming the Boltzmann case the Peltier term can be simplified to

$$h_{\text{Peltier}} = \frac{k_B T}{e} \cdot \left[\frac{j_{n,y} n}{N_c} \vec{\nabla} \left(\frac{N_c}{n} \right) - \frac{j_{p,y} p}{N_v} \vec{\nabla} \left(\frac{N_v}{p} \right) \right]. \tag{4.18}$$

For every such interface i a areal heat contribution of

$$\begin{aligned} H_{\text{Peltier,i}}\Big|_{y^-}^{y^+} &= \int_{y^-}^{y^+} h_{\text{Peltier}} dy \\ &= \frac{k_B T j}{e} \left(\Big[P_p \Big]_{y^- \in \text{p-layer}}^{y^+ \in \text{p-layer}} - \Big[P_n \Big]_{y^- \in \text{n-layer}}^{y^+ \in \text{n-layer}} \right) \\ &= \frac{k_B T j}{e} \left(\left[\ln \left(\frac{N_v}{p} \right) \right]_{y^- \in \text{p-layer}}^{y^+ \in \text{p-layer}} - \left[\ln \left(\frac{N_c}{n} \right) \right]_{y^- \in \text{n-layer}}^{y^+ \in \text{n-layer}} \right) \end{aligned} \tag{4.19}$$

is gained, where a constant vertical current density through every hetero surface $j_{n,y}|_{y \in \text{p-layer}} = j_{p,y}|_{y \in \text{n-layer}} = 0$ and $j_{n,y}|_{y \in \text{n-layer}} = j_{p,y}|_{y \in \text{p-layer}} = -j$.

Under the assumption of (3.5) and neglect of vertical carrier leakage, the electron and hole current density in the n- and p-doped layers is given by the doping concentration, $n \approx n_A$ and $p \approx -p_D$ with $n_A > 0$ and $p_D < 0$, respectively. The effective density of states N_c and N_v is calculated according to (3.10) with the corresponding effective masses m_n^* and m_p^* of the materials. In the active region $N_c \approx 10^{24}$ m^{-3} and $N_v \approx 5 \cdot 10^{24}$ m^{-3}, and an excess carrier density of $N = 3 \cdot 10^{24}$ m^{-3} is assumed. The Seebeck coefficients for electrons and holes P_n and P_p calculated in the Boltzmann limit are displayed in Fig. 4.1 for structure II and a current density of $j = 5 \cdot 10^7$ A/m^2. In the n- and p-doped layers electrons and holes absorb energy from the lattice while passing the gradients of the Seebeck coefficients and thus cool each interface under forward bias. At the boundary of n- and p-doped layers to the active region they release energy so that here a heating term occurs.

As these terms only exceed absolute values larger than 10^{15} W/m^3 at layer boundaries and partly compensate themselves, integrating the Peltier heat source density over the vertical direction yields $H_{\text{Peltier}} = \int h_{\text{Peltier}} dy = \sum_i H_{\text{Peltier,i}}|_{y^-}^{y^+} = -1$ MW/m^2. The absolute value of H_{Peltier} is much smaller than the areal heat source density of other heat sources, see Fig. 4.4, and accordingly the error made by neglecting this heat contribution is certainly negligible under CW operation.

Under pulsed operation, the impact could be higher, because the heat production is highest near the active region where the guided mode is localized. In Fig. 4.6(a) in solid black the temperature distribution is shown for laser operation with a 10 ns long pulse at an injection current of 150 A including vertical heat flow. Repeating the same estimation including Thompson-Peltier heat for a current density of $j = 3.75 \cdot 10^8$ A/m^2 appropriate for the investigated laser, results in an active region temperature increase from 5 K to 6.3 K after the 10 ns long pulse. For the employed model the resulting temperature induced refractive index change is $4.8 \cdot 10^{-4}$ after the 10 ns long pulse, displayed in Fig 4.6(b) as red dashed line. Including Thompson-Peltier heat and vertical heat flow the value is only slightly higher resulting in a thermally induced index change of $5.4 \cdot 10^{-4}$ after the 10 ns long pulse. Accordingly the integration of the Thompson-Peltier heat source into the model results only in a small improvement compared to the effort needed for its self-consistent inclusion.

4.2.3 Heat sources for the longitudinal-lateral approximate equations

We will now have a closer look into the longitudinal-lateral approximated equations for the heat source density, their distribution within the laser device as well as the relative share of the different heat sources to the total heating. The approximated heat source densities are gained by inserting the Ansatz (2.20), assuming that all spontaneously emitted radiation is instantaneously absorbed, neglecting the Seebeck coefficients and assuming an infinitly high conductivity in the n-doped layer.

The vertical heat source distribution is displayed in Fig. 4.2(a) and (b) for a 100 µm broad and 4 mm long laser (structure II Tab. A.3 and A.4) at a total injection current of 20 A and 150 A, respectively. The simulation time was $\tau_{\text{sim}} = 10$ ns and over the last $\tau_{\text{av}} = 3$ ns of that time span data was averaged. No index-guiding trenches were etched and the structure was considered to be implanted next to the injection stripe down to the vicinity of the active region ($d_{\text{res,imp}} = 0$ nm in Fig. 3.3). Any thermal effects on waveguiding and parameter dependence on temperature are neglected, to also discuss laser operation even at high injection currents, similar to pulsed operation.

The Joule heat is

$$h_{\text{J}} = j_p^2/\sigma_p \tag{4.20}$$

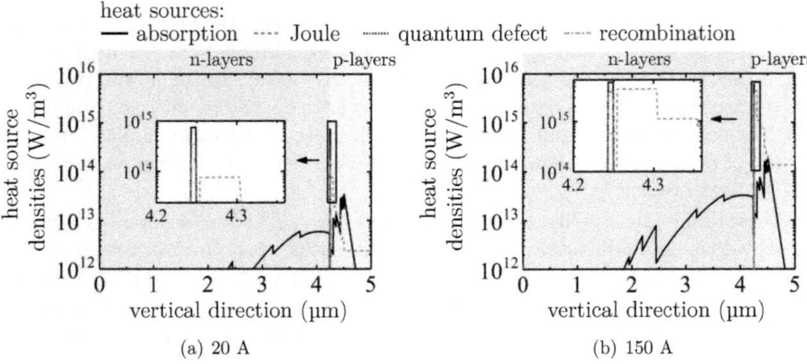

Figure 4.2: Time averaged absorption, Joule, recombination and quantum defect heat source densities as function of the vertical y-direction in the middle of the injection stripe for (a) 20 A and (b) 150 A injection current. Thermal effects on waveguiding and parameter dependence on temperature are neglected.

in the p-doped region and in accordance with the current flow model it is neglected in the n-doped region due to the very high electron conductivity. It is highest in the p-doped confinement layers due to their comparatively low conductivity, Fig. 4.2.

The heat source due to absorption of stimulated emitted photons is given by inserting (2.20) into (4.12) resulting in

$$h_{\text{abs}} = \begin{cases} v_g \hbar\omega d\alpha_0(y)\frac{n_r(y)}{\bar{n}}|\phi(y)|^2\|u(x,z,t)\|^2 \\ +(v_g\hbar\omega d)^2\beta(y)\frac{n_r}{\bar{n}}|\phi(y)|^4(\|u(x,z,t)\|^4+4|u^+(x,z,t)|^2|u^-(x,z,t)|^2) \\ \text{for } y \notin \text{active region,} \\ v_g\hbar\omega\Gamma f_N N\|u(x,z,t)\|^2 \text{ for } y\in \text{active region} \end{cases} \tag{4.21}$$

where $n_r(y)$, $\alpha_0(y)$ and $\beta(y)$ denote the vertical distribution of the real part of the refractive index, the absorption and the two-photon absorption coefficient, respectively, for which (2.33) and (2.36) must hold to ensure energy conservation. For a list of step-wise constant parameters for structure II see Tab. A.4.

The first term in Eq. (4.21) accounts for background absorption due to doping, the second term denotes two-photon absorption in the cladding and confinement layers and the third term is due to free carrier absorption in the active region. The main contributions arise in the active layer located at $y \approx 4.25$ µm due to the high non-equilibrium electron and hole densities and in the adjacent p- and n-doped confinement layers where the optical mode resides, Fig. 4.2.

The third source term is recombination heat in the active region, gained from (4.10) neglecting the Seebeck coefficients ,

$$h_{\text{rec}} = e\varphi_F(N,T)(AN+BN^2+CN^3). \tag{4.22}$$

Here it is assumed, that all spontaneously emitted photons are transferred into heat within the active region as has been discussed before. It only has a contribution in the active region at $y \approx 4.25$ µm as vertical carrier leakage is neglected, Fig. 4.2.

The last term taken into account denotes quantum defect heat generated in the active region as a result of an incomplete energy transfer from the carrier reservoir to

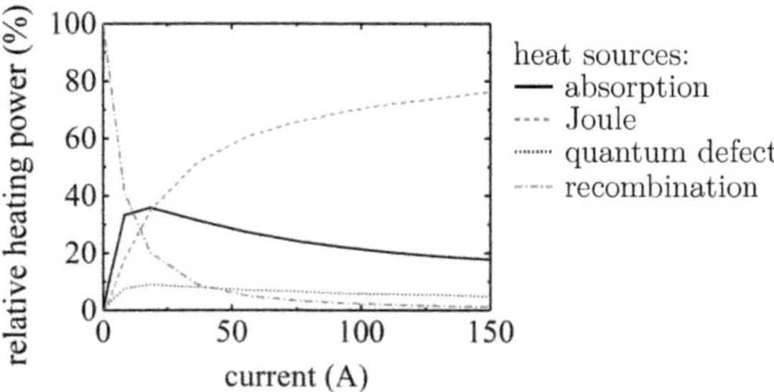

Figure 4.3: Relative heating powers as function of total injection current for the different heat contributions. Thermal effects on waveguiding and parameter dependence on temperature are neglected.

the radiation field and is gained from (4.11) neglecting the Seebeck coefficients and given as

$$h_{\text{defect}} = (e\varphi_F(N,T) - \hbar\omega)R_{\text{stim}}(N, \|u\|^2). \tag{4.23}$$

Accordingly, in the employed model the total heat source density is given by the sum over all heat sources h_i

$$h = \sum_i h_i = h_{\text{J}} + h_{\text{abs}} + h_{\text{rec}} + h_{\text{defect}} \tag{4.24}$$

with h_{J}, h_{abs}, h_{rec} and h_{defect} from (4.20) (4.21) (4.22) and (4.23), respectively.

In Figs. 4.2(a) and (b) it can be already seen that the share of the different heat sources changes relative to each other. This becomes even clearer in Fig. 4.3, where the relative heating powers of absorption, Joule, quantum defect and recombination heat to the total heating power $\int_V h_i dV / \int_V h dV$ are displayed as function of the total injection current for the same structure and simulation parameters as above. Near threshold most heat is generated due to non-radiative and spontaneous radiative recombination heating. With rising current the relative contribution of other heat sources increases so that at approximately 20 A absorption and Joule heating contribute to a similar amount. As the current rises further, Joule heating becomes more and more pronounced due to its quadratic dependence on the current, so that at a total injection current of 150 A it amounts to nearly 80 % of the total heat source.

In Fig. 4.4 the lateral-longitudinal distribution of the time averaged areal heat source densities $H = \int h dy$ including thermal waveguiding under CW operation at a total injection current of 20 A are displayed. For this simulation the same structure as before is used, however, the procedure as described in Section 4.5 is performed to obtain the thermally induced waveguide under CW operation. In total seven iterations were performed and for each iteration the simulation time was $\tau_{\text{sim}} = 8$ ns, whereas over the last $\tau_{\text{av}} = 3$ ns of that time span data was averaged.

At the displayed operation current absorption and Joule heating have the highest contribution to the total heating. Due to the high asymmetry in the facet reflectivity values both heating terms have higher values at the front $z = L$ compared to the rear facet $z = 0$. The result is a z-dependent temperature and accordingly a longitudinally varying temperature induced index profile. A similar result has also been obtained in [17] using stationary heat sources. The longitudinally varying temperature profile has a

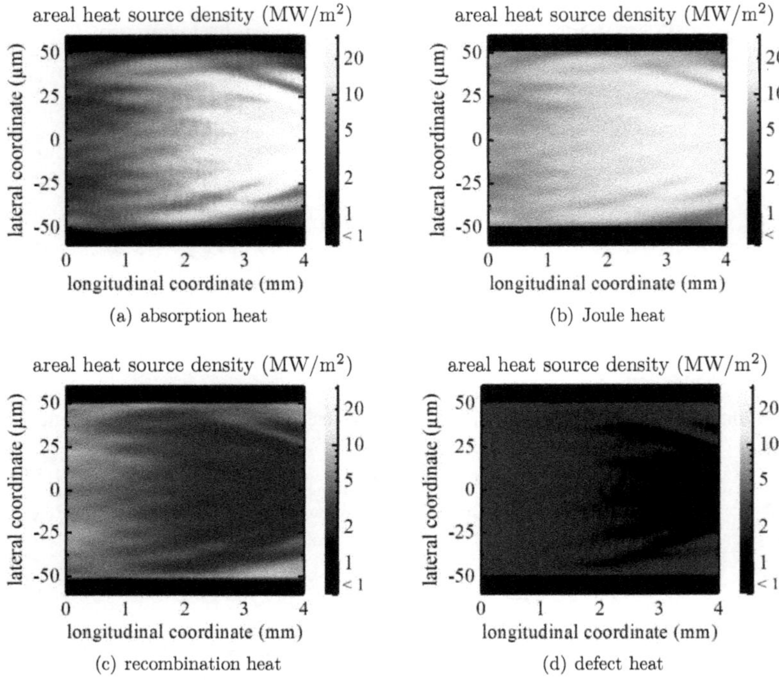

(a) absorption heat

(b) Joule heat

(c) recombination heat

(d) defect heat

Figure 4.4: Distribution of the time-averaged (a) absorption, (b) Joule, (c) recombination, and (d) quantum defect areal heat source densities $H = \int h dy$ as function of the lateral and longitudinal (x, z)-coordinate for CW operation at $I = 20$ A ($P_{\text{out}} \approx 19$ W). Thermal waveguiding is included in the calculations by performing the iterative procedure described in Section 4.5.

significant influence on the optical field and results in a narrowed optical field at the front facet, which is discussed in more detail in Section 6.4.2.

4.3 Energy conservation

In the previous section several approximations regarding the heat sources have been made. To ensure that the energy is still conserved for this simplified model, in the following energy conservation of the heat-flow equation (4.1) with the longitudinal-lateral approximated heat sources (4.20), (4.21), (4.22) and (4.23) is proven.

Reversing the rescaling of (2.20) and integrating over the device volume gives

$$\begin{aligned}\int_V c_{\mathrm{h}}\partial_t T\,dV - \int_V \vec{\nabla}(\kappa_{\mathrm{L}}\vec{\nabla}T)dV &= \int_V \frac{\vec{j}_p^{\,2}}{\sigma_p}dV + \int_V \frac{\varepsilon_0 c n_{\mathrm{r}}}{2}\alpha\|E\|^2 \\ &+ \int_V \left(\frac{\varepsilon_0 c}{2}\right)^2 \bar{n} n_{\mathrm{r}}\beta(\|E\|^4 + 4|E^+|^2|E^-|^2)dV \\ &+ \int_V e\varphi_F(R_{\text{non-rad}} + R_{\text{sp}})dV + \int_V [e\varphi_F - \hbar\omega]R_{\text{stim}}dV. \quad (4.25)\end{aligned}$$

Inserting (2.16) and (3.14), assuming that the spontaneous emission into the lasing modes is negligible (as lasing operation far above threshold is considered) and that the spontaneous radiation into all other modes is instantaneously reabsorbed in the cavity leads to

$$\begin{aligned}\int_V c_{\mathrm{h}}\partial_t T\,dV - \int_V \vec{\nabla}(\kappa_{\mathrm{L}}\vec{\nabla}T)dV &= \int_V -\vec{j}_p\vec{\nabla}\varphi_p dV + \int_V e\varphi_F(R_{\text{non-rad}} + R_{\text{sp}} + R_{\text{stim}})\,dV \\ &- \partial_t \int_V u_{\text{rad}}\,dV - \int_V \nabla\cdot S_{\text{rad}}\,dV. \quad (4.26)\end{aligned}$$

Furthermore using Green's identity,

$$\int_V \nabla\phi\nabla\psi dV = \int_{\partial V} \phi\nabla\psi\cdot\vec{n}dS - \int_V \phi\nabla^2\psi dV, \quad (4.27)$$

where ∂V denotes the surface of the device with $\vec{n}$ being the normal vector, yields

$$\int_V c_{\mathrm{h}}\partial_t T\,dV - \int_{\partial V} \kappa_{\mathrm{L}}\vec{\nabla}T\cdot\vec{n}dS = -\int_{\partial V} \varphi_p\vec{j}_p\cdot\vec{n}dS + \int_V \varphi_p \underbrace{\vec{\nabla}\cdot\vec{j}_p}_{=-e(R+\partial_t N)} dV, + \int_V e\varphi_F R\,dV \quad (4.28)$$

$$- \partial_t \int_V u_{\text{rad}}\,dV - \int_V \nabla\cdot S_{\text{rad}}\,dV \quad (4.29)$$

In the active region where $R \neq 0$ the hole potential is given by the Fermi-potential $\varphi_p = \varphi_F$ and hence the terms containing the recombination rate cancel, and we gain

$$\begin{aligned}\int_V c_{\mathrm{h}}\partial_t T\,dV - \int_{\partial V} \kappa_{\mathrm{L}}\vec{\nabla}T\cdot\vec{n}dS &= -\int_{\partial V} \varphi_p\vec{j}_p\cdot\vec{n}dS - e\int_V \varphi_F\partial_t N\,dV \\ &- \partial_t \int_V u_{\text{rad}}\,dV - \int_V \nabla\cdot S_{\text{rad}}\,dV. \quad (4.30)\end{aligned}$$

In what follows the steady state is considered ($\partial_t = 0$) and no flow of electrical current through the surface is assumed, except at the p-contact at $y = H$:

$$\vec{j}_p\cdot\vec{n} = 0 \quad \text{for} \quad y \neq H. \quad (4.31)$$

Between the contacts a forward bias U is applied, so that

$$\varphi_p|_{y=H} = U \qquad \text{and} \qquad \varphi_p|_{y=0} = 0 \quad (4.32)$$

hold. The normal components of the hole current density at the n-contact $y = 0$ are assumed to vanish,

$$j_{p,y}|_{y=0} = 0. \quad (4.33)$$

Similarly, no heat flow through the surface, except at the surface located at $y = H$, attached to the heatsink is assumed, where

$$\kappa_{\mathrm{L}} \partial_y T|_{y=H} = \frac{T_{\mathrm{ref}} - T|_{y=H}}{r_{\mathrm{th}}} \tag{4.34}$$

with r_{th} being the thermal transmission resistance (unit Km^2/W). By inserting the boundary conditions into (4.30) we get for the steady state

$$\iint \frac{T|_{y=H} - T_{\mathrm{ref}}}{r_{\mathrm{th}}} \, dx dz = UI - \int_V \nabla \cdot S_{\mathrm{rad}} \, dV \tag{4.35}$$

with $j = -j_{p,y}|_{y=H}$ and $I = \iint j \, dx dz$. The lhs of (4.35) is the heat flow to the heatsink. The first term on the rhs is the electric input power UI and the second term is the optical power that leaves the cavity. With the boundary conditions (2.10) and (2.11), the second term, *cf.* Eq. (2.15), can be shown to be

$$\int_V \nabla \cdot S_{\mathrm{rad}} \, dV = P_0 + P_{\mathrm{L}}. \tag{4.36}$$

Thus in the steady state the total heat generated in the cavity is given by the electrical input power UI minus the optical output power.

4.4 Treatment of pulsed operation (no-heat-flow approximation)

Under pulsed operation self-heating is usually neglected because thermal build-up times are much longer than the pulse lengths. However, the heat is generated near to the active layer in the same region where the guided wave is localized. As this region is small and its thermal build-up time is much shorter than that of the whole device, short-time local heating can be expected to influence the optical pulse formation and should be considered.

Due to the fluctuation of field intensity and carrier density, the heat sources and the resulting temperature distribution pattern are non-stationary. It is, however, unreasonable to solve the heat-flow equation (4.1) over tens of microseconds with the sub-picosecond temporal resolution of the opto-electronic model in the large spatial domain sketched in Fig. 1.1(d). Fortunately, inherent traits of heat generation exist that make it possible to simplify the problem considerably. They will be discussed in this section to deduce an easily integrable equation for the thermally induced index change under short pulse operation.

Most of the heat is generated in the active layer or very close to it, Fig. 4.2. Here the vertical temperature gradient is highest. Assuming constant c_h and κ_L, the heat-flow equation (4.1) in only one dimension is given by

$$\frac{\partial T}{\partial t'} - \frac{\partial^2 T}{\partial y'^2} = h(\kappa_L y', c_h t') \tag{4.37}$$

with the substitutions $t = c_h t', y = \sqrt{\kappa_L} y'$. For initially $T = T_{\mathrm{HS}}$, the solution is [72]

$$T(y', t') = \int_{-\infty}^{\infty} d\xi' \int_0^{t'} ds' h(\sqrt{\kappa_L}\xi', c_h s') \Psi(y' - \xi', t' - s') + T_{\mathrm{HS}} \tag{4.38}$$

with the fundamental solution

$$\Psi(\Delta y', \Delta t') = (4\pi\Delta t')^{-1/2} \exp\left\{-\frac{\Delta y'^2}{4\Delta t'}\right\}. \tag{4.39}$$

Ψ is normalized to unity for any $\Delta t' > 0$ and drops rapidly beyond the radius $\Delta y'^2 = \Delta t'$, Fig. 4.5. With dimensions, this yields the definition of thermal diffusivity α_D which gives the ratio between radius of the spreading heat sphere l^2 and elapsed time τ,

$$\alpha_D = \frac{\kappa_L}{c_h} = \frac{l^2}{\tau}. \tag{4.40}$$

Thus thermal diffusivity measures how well a substance conducts heat relative to its capacity to store it. By inserting the constant coefficients $c_h = 1.7 \cdot 10^6$ WsK^{-1}m^{-3} and $\kappa_L = 12$ W(Km)$^{-1}$ obtained as mean values of active and surrounding layers, *cf.* Tab. A.4, into Eq. (4.40) a diffusivity of

$$\alpha_D \approx 10^{-5}\ \mathrm{m^2/s} \approx \frac{(100\ \mathrm{nm})^2}{1\ \mathrm{ns}}, \tag{4.41}$$

is obtained. Accordingly, the generated heat during a 1 ns simulation interval is spread by only about 100 nm and transverse heat flow is certainly negligible because this distance is much smaller than the transverse inhomogeneity of the heat source density, Fig. 4.4.

If the heat source h is time independent, the temporal integral in (4.38) can be performed analytically, yielding

$$T(y', t') = \sqrt{t'} \int_{-\infty}^{\infty} d\xi'\ h(\sqrt{\kappa_L}(y' + \xi'))w\left(\frac{\xi'^2}{4t'}\right). \tag{4.42}$$

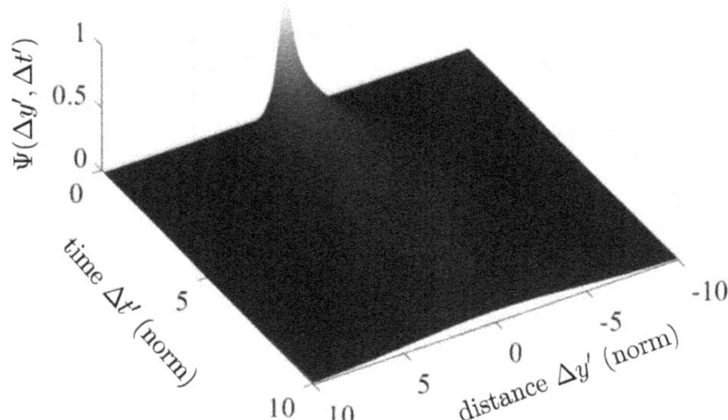

Figure 4.5: The fundamental solution of the heat equation $\Psi(\Delta y', \Delta t')$ as function of normalized time $\Delta t'$ and distance $\Delta y'$.

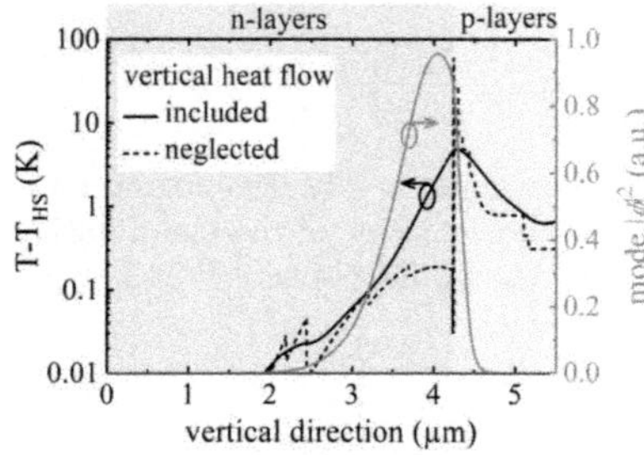

(a) Black left axis: Vertical temperature distributions at the end of the 10 ns pulse including vertical heat flow (black solid) derived from Eq. (4.43), and neglecting heat flow (black dashed). Red right axis: mode profile $|\phi(y)|^2$.

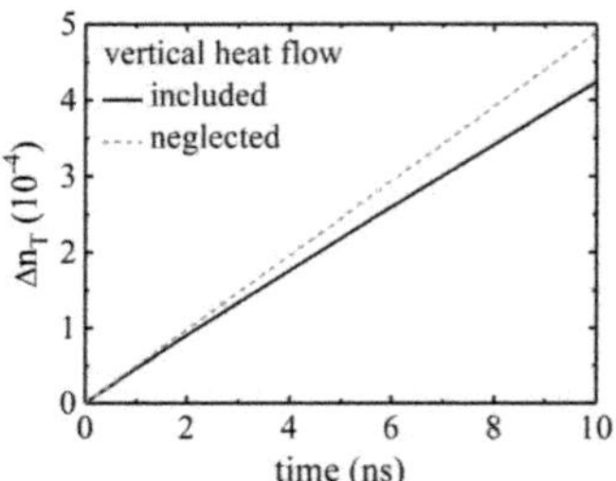

(b) Thermal contribution to the refractive index, Eq. (2.30), versus time including vertical heat flow (black solid) with $T(y,t)$ from Eq. (4.43), and neglecting heat flow (black dashed) from Eq. (4.44).

Figure 4.6: Impact of vertical heat flow for a 10 ns long 150 A pulse in the stripe middle. Constant coefficients $c_h = 1.7 \cdot 10^6$ WsK^{-1}m^{-3} and $\kappa_L = 12$ W(Km)$^{-1}$ are assumed.

Resubstituting $t' = t/c_h$, $y' = y/\sqrt{\kappa_L}$ and $\xi' = \xi/\sqrt{\kappa_L}$, gives

$$T(y,t) = \frac{1}{\sqrt{c_h \kappa_L}} \int_{-\infty}^{\infty} d\xi \; h(y+\xi)\; \sqrt{t}\; w\left(\frac{c_h \xi^2}{4\kappa_L t}\right), \qquad (4.43)$$
$$\text{with } w(q) = \left\{ \sqrt{q}\left[\operatorname{erf}(\sqrt{q}) - 1\right] + \frac{1}{\sqrt{\pi}} e^{-q} \right\}.$$

After a 10 ns pulse the vertical temperature distribution according to Eq. (4.43) is displayed in Fig. 4.6(a) (left axis - black solid) in the stripe middle. A vertically infinite domain with homogeneous material and constant coefficients $c_h = 1.7 \cdot 10^6$ WsK^{-1}m^{-3} and $\kappa_L = 12$ W(Km)$^{-1}$ is assumed. The fluctuating heat source density h at each position is replaced by its average along the whole pulse, which is displayed in Fig. 4.2(b). Moreover, heat flow in the longitudinal-lateral (x, z)-plane is disregarded. This is justified because the extensions of the lateral-longitudinal inhomogeneity in the heat source densities of the order of tens respective hundreds of micrometers (*cf.* Fig. 4.4) as well as the vertical size of the device are much larger than the expected heat spreading in the range of some hundred nanometers.

If vertical heat flow were negligible the temperature could be retrieved simply by integrating Eq. (4.1) with $\kappa_L = 0$. The result is displayed in Fig. 4.6(a) (left axis - black dashed) for comparison. Significant differences in the temperature distributions are clearly visible. Within the active region for instance a very high maximum temperature of 60 K is reached when vertical heat flow is neglected. However, when vertical heat flow is included a temperature of only 5 K is reached, because the temperature distribution is smoothed out considerably. Therefore, all parameters of the model including occupation probabilities are derived for the heat-sink temperature $T = T_{\text{HS}}$. This limits the no-heat-flow approximation to short pulses.

However, the main interest lies not in the actual temperature distribution, but primarily in the thermally induced refractive index change (2.30) and to obtain it, the vertical temperature distribution needs to be multiplied with the differential index n'_T

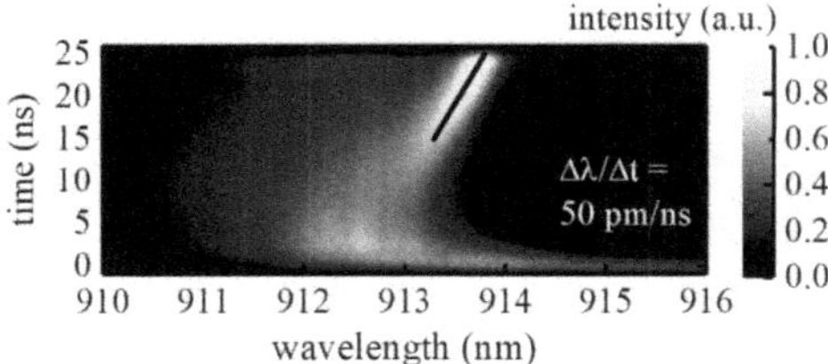

Figure 4.7: Measured time evolution of the emission spectra of a laser operated with 25 ns long pulses at an injection current of 6.3 A and repetition rate of 10 kHz (measured with a streak camera and averaged over 100 shots).* The black line indicates a linear fit of the maximum wavelength as function of time with the wavelength shift $\Delta\lambda/\Delta t = 50$ pm/ns.

and weighted with the vertically wide mode profile $|\phi(y)|^2$ shown in Fig. 4.6(a) on the right axis and integrated over the whole profile.

Thus, irrespective of the large local temperature differences obtained with and without vertical heat flow, the thermally induced index differs only marginally between the two cases, Fig. 4.6(b). Since high-power lasers generally use wide near fields to avoid damage due to extreme intensities, the no-heat-flow approximation is a generally reasonable approach to derive the thermal feedback on waveguiding for pulses of high power up to a length of at least 10 ns.

Heat flow is therefore negligible when simulating short transients. Differentiating (2.30) with respect to time and inserting (4.1) with $\kappa_L = 0$, the ordinary differential equation

$$\partial_t \Delta n_T = \int \frac{n_T'}{c_h} |\phi(y)|^2 h(N, \vec{j}, \|u\|^2) dy \tag{4.44}$$

for the thermally induced index is derived. At a given instant t, the actual distribution of local heat sources has to be inserted here. Accordingly, only one equation for the temporal variation of Δn_T has to be considered in each point (x, z) of the lateral-longitudinal plane. It is integrated much easier in each node of the spatial grid than the original partial differential heat-flow equation (4.1) .

4.4.1 Experimental validation

Experimental evidence of short time local heating is visible in Fig. 4.7 and can also be found in e.g. Ref. [73].* The lateral aperture of the experimentally investigated laser in Fig. 4.7 is defined by an etched mesa with a width of 100 µm. It is operated with a 25 ns long pulse and a low pulse repetition rate of 10 kHz in order to exclude heating between the pulses. Details on the experimental setup can be found in [74]. During the beginning of the pulse, when the active region is flooded with carriers, the emission spectrum shifts to shorter wavelengths. After relaxation the center wavelength starts shifting to longer wavelength as a result of self-heating in and around the active region.

The maximum wavelength shift is $\Delta\lambda/\Delta t = 50$ pm/ns. With $\Delta\lambda/\Delta T = 0.33$ nm/K known from independent measurements, this corresponds to a temperature increase of $T = 1.5$ K in and around the active region for a 10 ns long 6.3 A current pulse. For

*The author thanks H. Christopher and A. Klehr for the pulsed measurements.

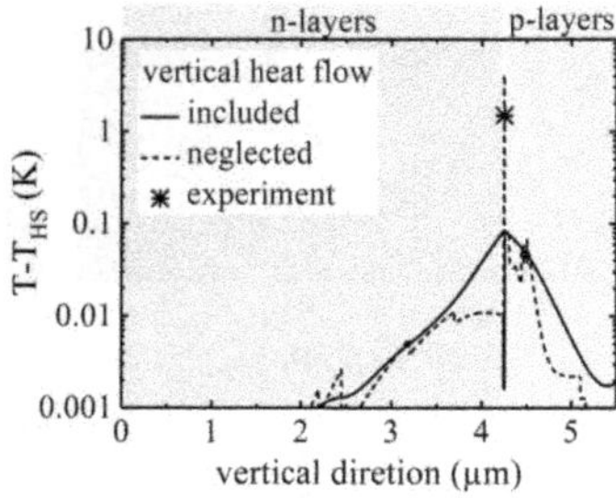

Figure 4.8: Simulated vertical temperature distributions at the end of a 10 ns long 6.3 A current pulse in the stripe middle including vertical heat flow (black solid) derived from Eq. (4.43), and neglecting heat flow (black dashed). The black star indicates the experimentally obtained temperature in and around the active region. Constant coefficients $c_h = 1.7 \cdot 10^6$ WsK^{-1}m^{-3} and $\kappa_L = 12$ W(Km)$^{-1}$ are assumed.

6.3 A the simulation yields temperature increases inside the active region of $T = 4.1$ K and $T = 0.1$ K without and with vertical heat flow, respectively (Fig. 4.8).

Thus, the experimental value lies below the temperature value obtained without heat flow but is by a factor 15 larger than the best estimate including vertical heat flow. As the used parameter values are known with significantly better precision, this suggests that the heat flow near the active region is inhibited by a mechanism not contained in the model.

Heat flow can for example be inhibited by longitudinal optical phonons with finite decay times that are generated via carrier relaxation [75, 76]. As the longitudinal-optical-phonon life time is in the picosecond range and thus much smaller than the considered pulse length of 10 ns, their impact can be neglected. A more probable cause is the thermal boundary resistance [77] of interfaces caused by reflection of acoustic phonons on hetero interfaces. Depending on the number of hetero-interfaces it could increase the temperature of the active region by some fractions of Kelvin [78]. Although the cause of inhibited heat transport remains unclear and is subject to further studies, it implies that the no-heat-flow approximation is even more realistic than expected.

4.5 Treatment of continuous-wave operation

We will now turn to CW operation. Here, the extremely long thermal build up can't be calculated with the no-heat-flow approximation. But in the later quasi-steady state, the rate of heat generation can be decomposed in a time-constant mean contribution $\bar{h}$ and a contribution $h_{\text{fluct}} = h - \bar{h}$ fluctuating around zero. h is the total instantaneous heat production (4.6). Accordingly, the heat-flow equation (4.1) is split into

$$0 = \nabla \kappa_L \nabla \bar{T} + \bar{h} \qquad \text{and} \tag{4.45}$$

$$c_h \frac{\partial T_{\text{fluct}}}{\partial t} = \nabla \kappa_L \nabla T_{\text{fluct}} + h_{\text{fluct}} \tag{4.46}$$

with boundary conditions following from (4.2). Obviously, the sum of the two temperatures obeys the full heat-flow equation (4.1).

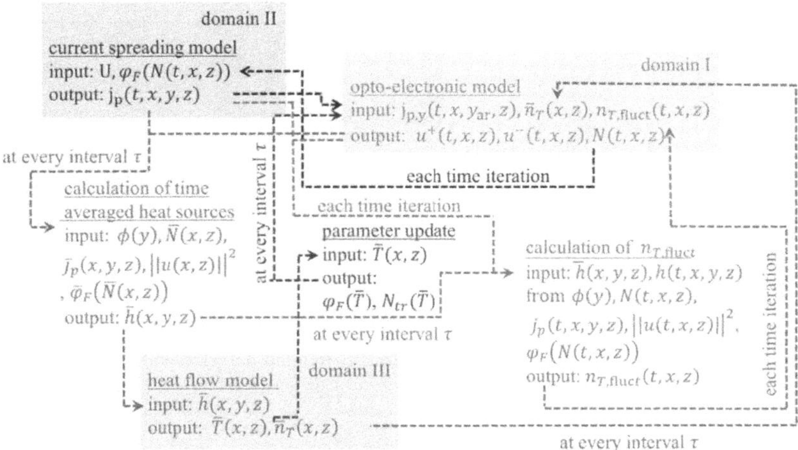

Figure 4.9: Flow chart of the coupling between opto-electronic, current-spreading and heat-flow model. The domain numbers correspond to the different simulation domains marked in Fig. 1.1

Treating the stationary equation (4.45) is challenging because heat flow dominates and, in general, it must be solved in the full 3-D domain including submout and substrate. However, due to longer characteristic distances in longitudinal direction the longitudinal heat-flow component is neglected and the static problem (4.45) is solved within multiple lateral-vertical (x-y) cross sections, see Fig. 1.1(d).

Furthermore, the mean heat production $\bar{h}(x,y,z)$ as a function of the coordinates is not known in advance. Therefore, an iterative approach is applied, which is sketched in a flow chart in Fig. 4.9. In the first step of the iteration the electro-optic model is solved together with the current spreading model under isothermal conditions with $T = T_{\text{HS}}$. The interaction of the two models is depicted in Fig. 4.9 with beige and blue colors. During this run, $\bar{h} = 0$ is set. In all following iterations, $\bar{h}$ is calculated as temporal average of the total heat production over the interval τ of the preceeding iteration. For the construction of the time averaged heat sources the vertical mode profile $\phi(y)$, time averaged photon density $\|\bar{u}(x,z)\|^2$, carrier density $\bar{N}(x,z)$ and accordingly Fermi-potential $\bar{\varphi}_F(\bar{N})$ from the electro-optic solver and the current density $\bar{j}(x,y,z)$ from the current spreading model are needed, see red section in Fig. 4.9. By solving (4.45) the time averaged distributions of temperature $\bar{T}(x,z)$ and thermally induced refractive index change $\bar{n}_T(x,z)$ are obtained.

The updated thermally induced refractive index change $\bar{n}_T$ enters the opto-electronic model for the next iteration. The updated stationary temperature $\bar{T}$ for each interval τ serves as reference for parameter values in the next iteration, see grey section in Fig. 4.9.

The time scale of the fluctuations $h_{\text{fluct}} = h - \bar{h}$ entering (4.46) is typically in the sub-nanosecond range. Thus, Eq. (4.46) is treated within the no-heat-flow approximation described above along a few-nanoseconds-long simulation interval τ, where the fluctuating part $h_{\text{fluct}} = h - \bar{h}$ is inserted into (4.44) to obtain the thermal impact on

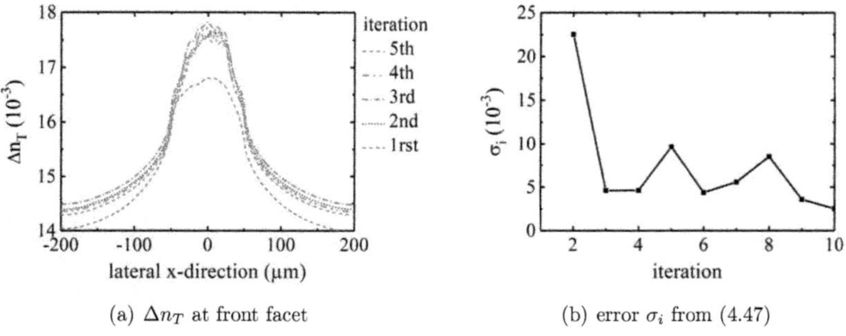

(a) Δn_T at front facet

(b) error σ_i from (4.47)

Figure 4.10: Evolution of the time-averaged thermally-induced refractive index change profiles Δn_T and error from iteration i to $i+1$ derived from Eq. (4.47) for 10 consecutive iterations.

wave-guiding $\Delta n_{\mathrm{T, fluct}}$. It is calculated from the time averaged heat source density $\bar{h}$ of the previous iteration and the actual heat source density h in every time instance, see green section in Fig. 4.9. Information on the numerical implementation, the spatial and temporal scales as well as the required simulation time can be found in Appendix D or in Ref. [44].

To check convergence the time-averaged thermally induced index $\Delta \bar{n}_T(x, z)$ is used as reference parameter. The error from iteration i to $i-1$ is defined as

$$\sigma_i = \sqrt{\frac{\iint \left[n_{T,i-1}(x,z) - n_{T,i}(x,z)\right]^2 dxdz}{\iint n_{T,i}^2(x,z)dxdz}}. \tag{4.47}$$

For the CW simulation discussed in this section, the time-averaged thermally induced index at the front facet is shown in Fig. 4.10(a) for the different iteration steps and

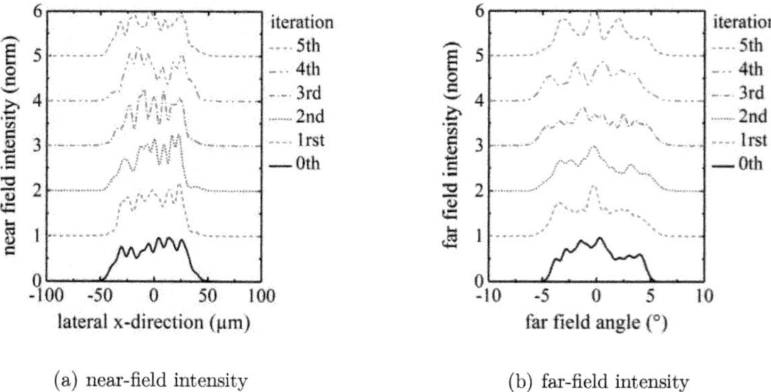

(a) near-field intensity

(b) far-field intensity

Figure 4.11: Profiles of the time-averaged (a) near and (b) far fields at the front facet for different iteration steps.

the errors σ_i in 4.10(b). In the 0th iteration the temperature was set to $\bar{T} = T_{\mathrm{HS}}$. Δn_T of the 1rst iteration is calculated from the time-averaged heat sources of the 0th iteration. The corresponding near and far fields of the different iteration steps are displayed in Fig. 4.11. For the 0th iteration where $\Delta\bar{n}_T = 0$ the far-field divergence is smaller compared to the other iterations. Furthermore the profile of $\Delta\bar{n}_T$ is flatter and consequently σ_i is higher compared to the other iteration steps. However $\Delta n_T(\bar{T})$ converges rapidly afterwards, whereas weak fluctuations around a constant level remain. These can be attributed to the chaotic spatio-temporal dynamics also visible in Fig. 4.11.

4.6 Summary

In this chapter a physically realistic and yet numerically applicable thermal model for BA lasers is presented, that accounts for slow and fast temperature contributions. The model bases on [45, 46] and heat sources due to absorption, Joule, recombination and defect heating are taken into account. They are formulated here in accordance with the optical longitudinal-lateral approximate equations and the carrier transport model presented in Chapter 2 and Chapter 3.

Further approximations are made to the energy-transport model, particularly vanishing Seebeck coefficients are assumed in agreement with the electronic model and accordingly Thompson-Peltier heating is neglected. As the Peltier heat terms only exceed absolute values larger than 10^{15} W/m^3 at layer boundaries and partly compensate themselves due to different signs in the contributions, they are negligible for the calculation of self-heating under CW and pulsed operation.

At moderate injection currents the calculated heat sources are highest in the active region due to a high contribution of recombination and absorption heating. For pulsed operation with currents far above threshold Joule heating in the p-doped layers is highest due to its quadratic dependence on the injection current. For a laser with asymmetric facet reflectivity values operated far above threshold, heating is higher at the front facet compared to the rear facet due to high contributions of absorption and Joule heating. The resulting influence on wave-propagation is discussed in more detail in Section 6.4.2. Furthermore energy conservation is shown for the employed equations.

Under short pulse operation with pulses in range of some nanoseconds, the temperature dependence of simulation parameters will be neglected. As the heat is mainly generated near the active region where the optical mode resides, temperature has an impact on waveguiding and is considered. To obtain the thermally induced index change, the vertical temperature distribution needs to be multiplied with the differential index, weighted with the vertically wide mode profile and integrated over the vertical direction. Thus although the temperature differs strongly including heat flow or neglecting it, the thermally induced refractive index remains nearly unaffected. Since high-power lasers generally use wide near fields, this approach is generally reasonable to derive the thermal feedback on waveguiding under short pulse operation. A comparison of the active region temperature calculated by the bulk heat-flow equations with measured ones indicates an inhibited heat flow. The reason for this may be thermal boundary resistances of hetero interfaces. However, this implies that the no-heat-flow approximation is even better than anticipated.

CW operation is solved by an iterative coupling of the electro-optical models to the heat transport equations to obtain the thermally induced refractive index profile and updated temperature dependent parameters. The thermally induced refractive index profile converges rapidly within few iterations, whereas weak fluctuations around a constant level remain which can be attributed to the chaotic spatio-temporal dynamics in BA lasers.

Together with the opto-electronic models, this results in a time-dependent quasi-three-dimensional opto-electronic and thermal model, that describes well essential qualitative characteristics of real BA laser devices and can be used to analyze spatio-temporal phenomena in BA lasers as discussed in the next sections.

Chapter 5

Power saturation under short pulse operation

Power saturation refers to the reduction of the slope of the power-current (PI) characteristics with increased current. Different phenomena exist that increase the internal losses so that an increasing contribution of the injected current does not lead to stimulated emission and thus a bending of the PI curve is the result.

Under continuous-wave (CW) operation power saturation is mainly attributed to overall device heating [5], as is presented in Section 7.1.1 and can be seen in Fig. 7.4. In this section the focus lies on power saturation under pulse operation with current pulses in the nanosecond range combined with low repetition rates. In this case device heating is strongly reduced, so that thermally induced power rollover due to enhanced loss mechanisms can be neglected as discussed in this chapter at the end of Section 5.3. Still, especially towards high injection currents a bending of the power-current characteristics is visible which is attributed to non-thermal power saturation.

Under pulsed operation different causes of power saturation have been discussed in the past. Leakage currents are caused by the transport of carriers into regions without stimulated recombination and can be divided into vertical and lateral leakage currents. Vertical leakage currents [6] are minority currents, caused by the accumulation and subsequent recombination of mainly electrons in the p-doped optical confinement layers. The vertical structure can be appropriately designed to reduce vertical leakage currents, which is done for the previously mentioned LiDAR device [79]. Lateral leakage currents are caused by current spreading in the p-doped region and can be e.g. reduced by implantation or etching, *cf.* Sections 6.3.3 and 7.1.

Spatial hole burning is a result of the interplay of spatial depletion of carriers due to stimulated emission and the insufficiently fast transport of injected carriers into the depleted regions. It has been investigated in the past for the longitudinal [7] as well as the lateral [8, 9] direction. The impact of two-photon absorption on the power-current characeristics has for example been investigated in [10, 11]. Furthermore there are gain compression mechanisms [12][13] such as spectral hole burning and carrier heating [14]. Similarly to spatial hole burning spectral hole-burning arises due to the interplay of the depletion of the carriers at a certain energy corresponding to the lasing wavelength due to stimulated emission. At high power densities the finite intra-band relaxation times are insufficiently fast to fill the hole formed by the optical transitions. Carrier heating arises as a result of the increased temperature of the carrier distributions above the lattice temperature by the removal of "cold" carriers near the band-edges due to stimulated emission or by the transfer of carriers to high energies within the bands by free-carrier, inter-valence band and two-photon absorption.

In all cases, the average carrier density within the cavity is increased to fulfill the threshold condition, which causes an increase of non-radiative and spontaneous radiative recombination as well as raised internal losses. Consequently, there is an increased contribution to the injected current that does not lead to stimulated emission, resulting into a reduced slope of the power-current characteristics.

Preceding publications showed that identifying the dominant mechanism is extremely difficult, as it depends on the laser emission wavelength, design, and material. Furthermore the impact of these mainly nonlinear effects depends on each other. As an example, gain compression alone in otherwise linear models, i.e. in the absence of spatial hole burning, with constant absorption, and a linear dependence of the gain and non-radiative and spontaneous radiative recombination on the carrier density, leads to a reduced differential efficiency, but the linearity of the power-current characteristic is preserved. Only due to the interplay with other nonlinearities such as e.g. Auger recombination, a bending of the PI curve appears. Consequently, it is difficult to separate the influence of individual effects on power saturation.

The theoretical analysis of the device in Ref. [79] is done with the laser simulation tool WIAS-TeSCA [58] that solves a sophisticated model for the stationary isothermal laser operation. The model resolves in detail the vertical carrier transport as well as the vertical and longitudinal variation of power and carrier density in the active quantum well. It does not contain any temporal or lateral variations of the optical power. Furthermore all thermal effects were neglected as the pulse repetition time of 100 µs prevents thermal memory effects between subsequent pulses.

As high-power BA lasers exhibit strong lateral power variations as well as temporal power fluctuations (*cf.* the measurement in Fig 5.4(a)) it can be expected that the impact of all terms nonlinear in the power will be modified by these strong spatio-temporal power variations compared to those models working with averaged powers.

In the following the traveling-wave based model as described in Chapters 2 and 3 is used to investigate the impacts of spatial hole burning, lateral current spreading, two-photon absorption and gain compression on a 5 mm long Fabry Pérot (FP) laser. Those results are compared to the simulation of a 6 mm long DBR laser with a 1 mm long DBR section to assess the influence of the DBR grating on the output characteristics. In Section 5.2 the influence of spatio-temporal power variations is investigated by comparing the simulation of the 5 mm long FP laser using the traveling wave model with the results of Ref. [79] where the full drift diffusion model is solved but the temporal and lateral variations of the intensity are omitted. An assessment of additional effects not included in this model is given in Section 5.3, followed by a conclusion in Section 5.4.

5.1 Spatial hole burning, current spreading, two-photon absorption and gain compression

In this section simulation results of FP and DBR lasers are compared utilizing the traveling wave model. Both FP and DBR lasers have a 50 µm wide contact stripe and the same length of the gain section of $L_{\text{gain}} = 5$ mm. For lateral optical and current confinement an index-guiding trench is etched with an effective index step of

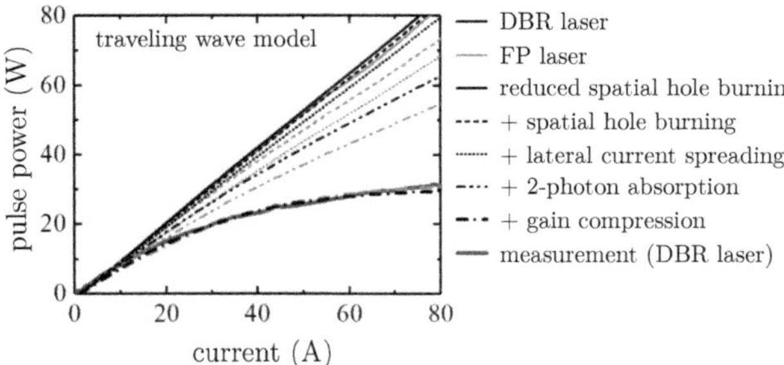

Figure 5.1: Influence of the DBR grating: Comparison of FP (black) and DBR laser (red) simulated with the traveling wave model with reduced spatial hole burning (continuous), plus spatial hole burning (dashed), plus two-photon absorption (dotted) and plus gain compression with $\epsilon_s = 2.4 \cdot 10^{-24}$ m^3 (dash dot). The experimental PI curve of a DBR laser operated in pulsed mode is also shown for comparison (continous gray).*

$\Delta n_0 = -3 \cdot 10^{-3}$ at $|x| \in [30:35]$ entering Eq. (2.28). The power reflectivity of the rear facet of the FP laser is $|r_0|^2 = 0.6$. The coupling coefficient of the Bragg grating has been adapted to provide the same reflectivity of $R_{\text{DBR}} \approx 0.6$ for a DBR grating length of $L_{\text{DBR}} = 1$ mm. Thus the coupling coefficient is $\kappa = 2000$ m^{-1} including losses of $\alpha_{\text{DBR}} = 900$ m^{-1} due to absorption and scattering. Further simulation parameters can be found in Tab. A.1.

For the calculation of the PI characteristics shown in Figs. 5.1 and 5.3 with the dynamic traveling-wave model, the procedure described in Section 2.4 is applied. The height of the staircase voltage ramp is 0.2 V or 0.004 V depending on the series resistivity r_s (3.46). The simulation time is $\tau_{\text{sim}} = 2$ ns, whereas over the last $\tau_{\text{av}} = 1$ ns of τ_{sim} the output power given by Eq. (2.44) is averaged.

The PI characteristics of DBR and FP lasers simulated with the traveling wave model are shown in Fig. 5.1. For comparison, the measured peak power versus peak current of a DBR laser operated by 10 ns long current pulses with a repetition frequency of 10 kHz is also shown.* For a series resistivity (3.46) of $r_s = 10^{-6}$ Ωcm^2 spatial hole burning is considered to be highly reduced, the actual value for the investigated structure is $r_s = 5.5 \cdot 10^{-5}$ Ωcm^2 ("with spatial hole burning"). The dependence of spatial hole burning on the series resistivity due to current self-distribution is described in more detail in the preceding Section 3.2.2.

When spatial hole burning, lateral current spreading, two-photon absorption and gain compression are omitted the slope efficiencies of DBR and FP lasers are nearly equivalent. Power reduction due to spatial hole burning is higher for the DBR laser compared to the FP laser, whereas two-photon absorption and gain compression influence the FP laser more. Accordingly, the PI curves of both lasers are almost identical and the measured PI curve of the DBR laser can be well reproduced with a high gain compression factor of $\epsilon_s = \Gamma \cdot 1.8 \cdot 10^{-22}$ m^3 $= 2.4 \cdot 10^{-24}$ m^3 entering the logarithmic gain model in Eq. (2.31). This is the same gain compression factor used in Ref. [79], which corresponds to a saturation power of $P_s = 4$ W.

---

*The author thanks H. Christopher and A. Klehr for the pulsed measurements.

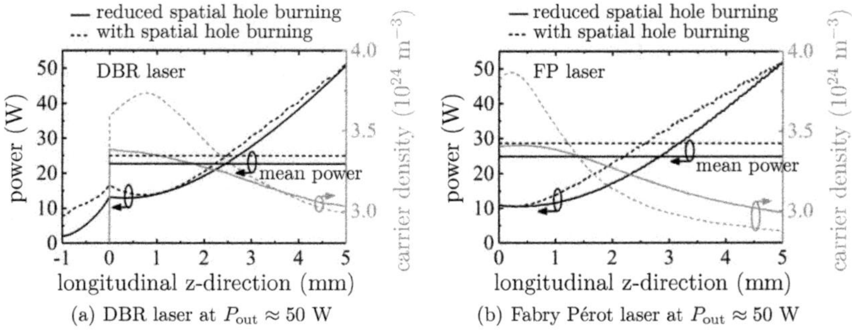

(a) DBR laser at $P_{\text{out}} \approx 50$ W (b) Fabry Pérot laser at $P_{\text{out}} \approx 50$ W

Figure 5.2: Longitudinal variation of the time averaged power $P^+ + P^-$ from (2.12) (black - left axis) and time averaged carrier density (red - right axis) within the cavity for a DBR and FP laser at an injection current of 60 A. Solid and dashed curves represent the cases with reduced spatial hole burning ($r_s = 10^{-6}$ Ωcm^2) and where spatial hole burning is included ($r_s = 5.5 \cdot 10^{-5}$ Ωcm^2), respectively.

Literature values of the gain compression factor for the description of spectral hole burning and carrier heating are in the range of $\epsilon_s = \Gamma \cdot 10^{-23}$ m^3 $= 1.3 \cdot 10^{-25}$ m^3. Note, that the gain compression factor has to be multiplied with the active region confinement factor Γ to be used in the effective model [49]. When assuming the literature value gain compression has no impact on the PI curve (not shown here). Thus with this artificial gain compression factor not only spectral hole burning and carrier heating are included but also all other remaining effects that lead to the discrepancy between simulation and measurement in Ref. [79].

It is important to note that at an output power of ≈ 15 W where high-power lasers are operated in CW mode non-thermal gain compression has only a very small impact. However at higher injection currents such as 60 A the pulse power is halved from 60 W to 30 W.

The impact of two-photon absorption is comparatively high for the investigated asymmetric vertical structure due to its high intensity in the n-doped confinement layers leading to a high overlap of the lasing mode with higher band gap materials. For more symmetric structures two-photon absorption is shown to be of less importance [80].

The different impact of the nonlinear effects on FP and DBR lasers can be explained by spatial hole burning when looking at Fig. 5.2 where the influence of spatial hole burning on the longitudinal power and carrier density variation in the FP and DBR laser is visible. The influence of spatial hole burning on the lateral field distribution is discussed in Section 6.3.3. Spatial hole burning acts differently on the DBR and FP lasers: At the same injection current of 60 A the power variation is nearly equivalent for DBR and FP lasers with reduced spatial hole burning (black solid line in Figs. 5.2(a) and (b)). When spatial hole burning is included a higher power loss through the DBR is visible, see Fig. 5.2(a) at $z = -1$ mm where the power is increased from 2 W to 8 W, whereas the power loss at the rear facet stays constant for the FP laser, see Fig. 5.2(b) at $z = 0$ mm. Furthermore, the power within the cavity of the FP laser is higher when spatial hole burning is considered (black dashed in Fig. 5.2(b)). As a result,

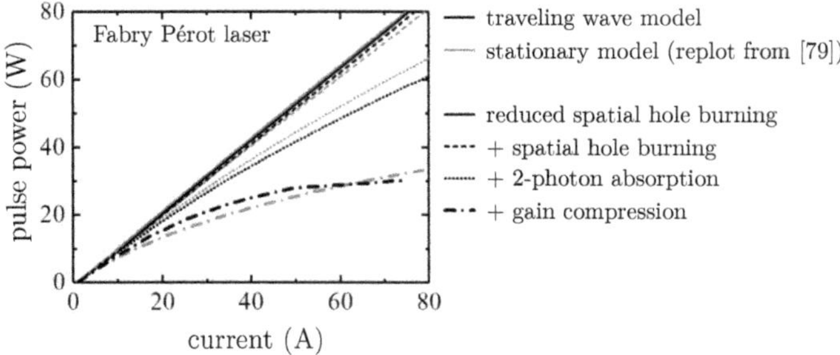

Figure 5.3: Model comparison: Simulated PI characteristics of a FP laser simulated with the traveling wave model (black) and with the stationary model from Ref. [79] (red) without non-linear effects (continous), plus spatial hole burning (dotted), plus two-photon absorption (dashed) and plus gain compression with $\epsilon_s = 2.4 \cdot 10^{-24}$ m^3 (dash dot).

two-photon absorption and nonlinear gain compression lead to higher losses for the FP laser compared to the DBR laser (Fig. 5.1) when spatial hole burning is accounted for.

5.2 Impact of spatio-temporal power variations

In Fig. 5.3 the simulated PI characteristics of the FP laser according to the traveling wave model as described in Chapters 2 and 3 is shown in black and the simulation results from Ref. [79] are replotted in red.

When spatial hole burning, two-photon absorption, and gain compression are omitted both simulated PI curves have the same slope. This shows that vertical carrier leakage is well restrained by the vertical epitaxial structure as discussed in Ref. [79] and that it is justified to use the simplified carrier transport model described in Chapter 3.

The inclusion of spatial hole burning alone has only a small influence on the output power in both cases. An additional inclusion of two-photon absorption has a smaller impact on the output power for the stationary simulation of Ref. [79] compared to the traveling-wave model studied here, which is a result of the spatio-temporal power variations. To estimate the influence of spatio-temporal power fluctuations on two-photon absorption and gain compression the photon density is exemplary studied at the output facet for an injection current of $I = 40$ A, Fig. 5.4(b). Here the mean photon density is $\langle||u(x,t)||\rangle_{x,t} = 2 \cdot 10^{24}$ m^{-3}, but photon density fluctuations temporarily reach much higher values. By calculating the two-photon absorption coefficient α_{2P} for the mean photon density, $\alpha_{2P} = f_{2P}\langle||u||^2\rangle_{x,t} = 36$ m^{-1} is obtained. However, by first including power fluctuations in the calculation and taking the time and spatial average afterwards the two-photon absorption coefficient reaches $\langle f_{2P} \cdot ||u||^4\rangle_{x,t}/\langle||u||^2\rangle_{x,t} = 55$ m^{-1} corresponding to a 50% increase. As a result, two-photon absorption is underestimated when stationary conditions are assumed.

After including gain compression using the artificial gain compression value of $\epsilon_s = 2.4 \cdot 10^{-24}$ cm^3, which was used earlier (Fig. 5.1) to fit the measured PI curve, both curves are almost identical. Including this phenomenological gain compression model has a lower impact on the PI curve when spatio-temporal fluctuations are

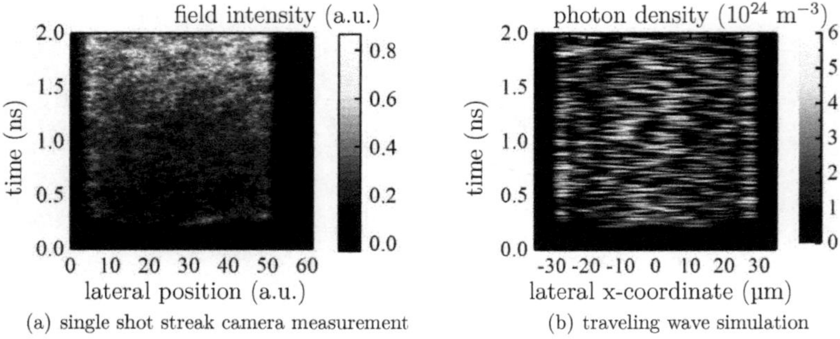

(a) single shot streak camera measurement

(b) traveling wave simulation

Figure 5.4: Comparison of (a) the field intensity of a single shot streak camera measurement (replot from [74]) and (b) the photon density from the traveling wave simulation (a moving mean over 10 ps was applied) at the front facet at a pulse peak output power of ≈ 23 W and current of 40 A.*

accounted for compared to the stationary case, Fig. 5.3. At first sight, this might seem surprising, because two-photon absorption has a higher impact on the output power when spatio-temporal phenomena are accounted for as discussed before. However, similarly to the fluctuation of the photon density the non-stationarity of the carrier density has to be considered as well. Calculating the temporal average of the unsaturated gain and dividing it by the temporal average of the saturated gain gives a factor of $\langle g' \ln(N/N_{\mathrm{tr}})\rangle_{x,t} / \langle g' \ln(N/N_{\mathrm{tr}})/(1 + \epsilon_{\mathrm{s}}\|u\|^2)\rangle_{x,t} = 4.2$. Whereas by directly calculating this factor from the mean photon density gives $1 + \epsilon_{\mathrm{s}}\langle \|u\|^2\rangle_{x,t} = 6$, which results in a 50 % greater factor for the decrease of the gain. As a result, similarly to two-photon absorption including spatio-temporal phenomena has also an impact on the output power when using the phenomenological gain compression model (2.31) and should not be neglected.

5.3 Estimation of additional effects not included in the model

In the presented model spatio-temporal fluctuations and power loss due to spatial hole burning, current spreading and two-photon absorption are self-consistently included. However, a reproduction of the measured PI curve can't be achieved, so that a phenomenological gain compression model is employed. Commonly, effects included by gain compression are spectral hole burning and carrier heating [12, 13]. However, the gain compression factor employed here is much higher which indicates that other effects play a role that have not been considered yet.

In Refs. [11, 10] indirect two-photon absorption due to free carrier absorption by two-photon absorption-created carriers is proposed to result in power saturation in pulsed high-power lasers. This effect is strongly dependent on the carrier lifetime and the effective drift-diffusion length. In most cases the resulting absorption coefficient can be neglected [80], except in assymetric structures as discussed here, where the

*The author thanks H. Christopher and A. Klehr for the pulsed measurements.

intensity peak of the vertical mode is not located at the position of the active layer [11]. Following the analysis of [11] the diffusion equation including recombination and carrier drain into the active region reads,

$$D_c \frac{\partial^2 N}{\partial y^2} + G_{2\mathrm{P}}(y) - R(N) = 0 \quad \text{with} \quad G_{2\mathrm{P}}(y) = \frac{\beta(y)}{\hbar\omega}\left(\frac{P}{W}\right)^2 |\phi(y)|^4 \tag{5.1}$$

with the diffusion constant $D_c = 2.6 \cdot 10^3$ m^2/s and the 2-photon-absorption carrier generation term $G_{2\mathrm{P}}(y)$ with the two-photon absorption coefficients from (B.3) and the normed mode profile $|\phi(y)|^2$. Properly incorporating Eq. (5.1) into the laser model lies beyond the scope of this thesis. However, a simple estimation of the two extreme cases where either the recombination or carrier diffusion is neglected, shall be discussed in the following for a power of $P = 30$ W and the laser stripe width $W = 50$ µm. By neglecting diffusion, i.e. setting $D_c = 0$ in Eq. (5.1), the carriers generated by two-photon absorption can be estimated by $N_{2\mathrm{P}} \approx G_{2\mathrm{P}} \cdot \tau$ with the minority carrier lifetime $\tau = 1$ ns, so that the effective absorption coefficient by the generated carriers is given by

$$\alpha_{2\mathrm{P,fc},1} = \frac{P^2\tau}{W^2\hbar\omega}(f_n + f_p)\int dy \beta_2(y)|\phi(y)|^6 \tag{5.2}$$

$$= 390 \text{ m}^{-1} \tag{5.3}$$

with the cross-section for free carrier absorption f_n and f_p for electrons and holes, respectively, yields an enormously high coefficient for the secondary two-photon absorption. By using Eq. (5.1), neglecting recombination $R(N) = 0$, whereas still considering drain into the active region, and twice integrating the resulting simplified carrier transport equation, yields the carriers that are generated by two-photon absorption,

$$N_{2\mathrm{P}}(y) = \frac{P^2}{D_c W^2 \hbar\omega}\int_0^y dy'' \int_{y''}^{\infty} dy' \beta_2(y')|\phi(y')|^4 \tag{5.4}$$

with the resulting absorption coefficient,

$$\alpha_{2\mathrm{P,fc},2} = (f_n + f_p)\int dy N_{2\mathrm{P}}(y)|\phi(y)|^2 \tag{5.5}$$

$$= 17 \text{ m}^{-1}. \tag{5.6}$$

Thus including only carrier diffusion reduces the previous value (5.3) by a factor of 20 to only 17 m^{-1}, identifying this effect as comparatively small for our device structure compared to the direct two-photon absorption coefficient. However, in order to gain a reliable assessment of the effect the full drift diffusion equations have to be solved.

By applying nanosecond-injection pulses with low repetition rates device heating is strongly reduced, as is visible in Fig. 4.6(a) of Section 4.4, where the maximum excess temperature reached within the active region is only 5 K for short-pulse operation with 10 ns long 150 A injection current pulses. As discussed in Section 4.4 under short pulse operation, where the no-heat-flow approximation is applied, all parameters of the model are derived for the heat-sink temperature $T = T_{\mathrm{HS}}$. Thermal waveguiding can be considered in the model, however, this has no impact on the power-current characteristics for currents up to 150 A as shown in Fig. 6.27(a) in Section 6.4 for a different vertical structure. To assess the influence of parameter changes with temperature, the FP laser simulation including spatial hole burning and current spreading (red dotted PI curve in

Fig. 5.1) is repeated at a heat sink temperature of $T_{\mathrm{HS}} = 305$ K instead of $T_{\mathrm{HS}} = 300$ K. The new parameter values are calculated according to the temperature dependence of parameters given in Appendix A and Tab. A.5. At 60 A injection current the output power reduces by only 0.6 A from 59.5 W, see red dotted PI curve in Fig. 5.1, to 58.9 W. Accordingly, the effect of device heating on the output power is negligible compared to other discussed power saturation mechanisms.

In lasers employing a QW active region, carriers belonging to the vertically confined bound QW states are not necessarily in thermal equilibrium with the carriers belonging to the higher unconfined free states, whereas for both ensembles different quasi-Fermi levels $E^{2D}_{F_{n,p}}$ and $E^{3D}_{F_{n,p}}$ can be assumed for the bound and free states, respectively. The capture-escape process between the bound and free states can be described by a net capture rate from the free states into the bound states, $R_n = R_{\mathrm{cap}} - R_{\mathrm{esc}}$. A simple expression for the net capture rate of the electrons is for example given in [81] as

$$R_n = \frac{n_{3D}}{\tau_n^c}\left[1 - \exp\left(\frac{E^{2D}_{F_n} - E^{3D}_{F_n}}{k_B T}\right)\right] \tag{5.7}$$

and depends on the free carrier density n_{3D}, on the Fermi-potentials of the bound $E^{2D}_{F_n}$ and free states $E^{3D}_{F_n}$, and is inversely proportional to the electron capture time τ_n^c. This capture-escape process influences the output power characteristics as also discussed in Ref. [81]. With increasing capture time τ_n^c the slope of the PI curve is reduced, because above threshold electrons accumulate outside the QW near the barrier resulting in higher losses. Depending on the capture time this reduction in the PI curve slope can be significant [81].

A further vertical carrier transport process not accounted for in the model, is the increase in vertical electron leakage currents with increasing bias due to the bending of the band edges [5]. To counteract minority carrier accumulation outside of the active region, the employed vertical epitaxial structure has a decreased thickness of the p-doped optical confinement layer [79], so that vertical leakage currents should have only a minor impact.

5.4 Conclusions

At high injection currents pulsed BA lasers experimentally exhibit a strong power saturation which is identified to be partly caused by spatial hole burning, current spreading and two-photon absorption whereas additional effects can be included in the theoretical model by a gain compression term.

Spatial hole burning influences the output power of DBR lasers and FP lasers of the same design differently. On the one hand inclusion of spatial hole burning leads to losses through the DBR resulting in power reduction. On the other hand it results in a higher intensity distribution within the cavity of a FP laser leading to higher losses due to two-photon absorption and gain compression. The power loss due to two-photon absorption is found to be higher when spatio-temporal power variations are included, whereas gain reduction by the phenomenological gain compression model is found to be smaller when spatio-temporal power variations are included. Both cases show, however, that spatio-temporal fluctuations have an impact on the output power and should be included in the analysis of power saturation.

Besides spectral hole burning and carrier heating additional effects that could influence the PI characteristics are indirect two-photon absorption and capture-escape processes which should be considered for further model expansions. The impact of two-photon absorption is comparatively high for the investigated asymmetric vertical structure due to its high intensity in the n-doped confinement layers leading to a high overlap of the lasing mode with higher band gap materials. For more symmetric structures two-photon absorption is shown to be of less importance [80].

Chapter 6

Factors influencing the lateral field profile

The multi-peaked and not diffraction-limited lateral field profile of wide-aperture semiconductor lasers has been a long standing problem and has been investigated in the past by numerous authors [2, 82, 83, 27, 84, 85, 86].

As shown in Ref. [47], a spontaneous break-up of the optical field into small filaments can happen in media with a focusing Kerr-nonlinearity. In semiconductor lasers a similar nonlinearity can be induced by spatial hole burning due to the dependence of the refractive index on the carrier density. Early results of simulations based on a Maxwell-Bloch type dynamic traveling-wave model with a sophisticated treatment of carrier kinetics supported this view [87]. For a hypothetical steady state an equation for this indirect Kerr-type nonlinearity can be formulated [27, 49].

An alternative understanding of the multi-peaked field structure of broad-area (BA) lasers bases on the simultaneous lasing of a large number of lateral waveguide modes [88]. Single mode emission becomes unstable just above threshold because, due to lateral spatial hole burning, any mode saturates the carrier density and consequently the modal gain in those parts of the active layer where the mode intensity is high. The modal gain in other parts rises with current, bringing more modes to threshold which can be additionally supported by a built-in or thermally induced waveguide. Indeed, recent experiments reveal, that even at currents several times above threshold the lateral modes can be clearly identified by spectrally-resolved near and far-field measurements [89, 16].

With increasing current a broadening of the near and far fields can be observed, which is commonly referred to as near or far-field "blooming". In the mode picture it can be understood by the broadening of the individual modes and the appearance of new modes with broader near and far fields [90].

In the absence of a thermally induced waveguide transverse instabilities arise due to lateral spatial hole burning and the nonlinear interaction between the electromagnetic field and the gain material. Concerning non-thermal blooming, the mode picture has been supported by a stationary model based on the incoherent superposition of multiple modes [21]. However, BA lasers exhibit an inherently non-stationary behavior and the assumption of stationary multimode conditions is questionable [54].

Thermal far-field blooming is well studied both experimentally [16] and theoretically [18, 19, 20] for continuous wave (CW) operation assuming a stationary temperature distribution. Recently a time-averaged longitudinally varying temperature profile was obtained with stationary temperatures [17]. Thermal waveguiding is usually neglected under pulsed operation because thermal build-up times up to milliseconds are much longer than the pulse lengths. However, the heat is generated near to the active layer in the same region where the guided wave is localized. This region is small and its thermal build-up time is much shorter than that of the whole device.

In the first two sections of this chapter the origins of transverse instabilities present in BA lasers are discussed in more detail. On the one hand, a simple theory for filamentation in the presence of a Kerr-nonlinearity is revisited, in Section 6.1, and compared to simulations of the traveling wave model to assess its significance for BA laser operation. On other hand, to see whether the multi-peaked and dynamic field structure obtained by the traveling-wave model can be understood by multi-mode lasing, in Section 6.2, the optical field is decomposed into lateral waveguide modes in a post-processing step. In Section 6.3 non-thermal far-field blooming and the dependence of transverse instabilities regarding parameters influencing the carrier reservoir (i.e. series resistivity (3.46), sheet resistance (3.47), and mobility) as well as the interaction between electromagnetic field and modal gain (i.e. differential index n'_N and gain g') are analyzed. In Section 6.4 the time-dependent approach to heating as described in Chapter 4 is used to study thermal waveguiding effects under pulsed and CW operation.

6.1 Modulation instability induced by Kerr nonlinearities

The transverse modulation instability formulated by Bespalov and Talanov [47], describes the breakup of a laser beam with a homogeneous intensity distribution into a beam with a spatially modulated intensity distribution. This phenomenon can happen in a medium with a focusing Kerr nonlinearity as a consequence of the growth of spatial irregularities initially present on the laser wavefront, which is sometimes referred to as "filamentation" [1]. Note, that the term modulation instability often refers to the temporal analogue, where a monochromatic laser field can become unstable to the growth of new frequency components [91]. This effect is, however, not subject of the discussion here.

In semiconductor lasers an indirect Kerr-type nonlinearity can be induced by the dependence of the refractive index on the carrier density because due to a depletion of the injected carrier density the refractive index increases in regions of high intensity leading to self-focusing [2]. For a hypothetical steady state an equation for the indirect Kerr-type coefficient can be formulated [27, 49].

In the following sections 6.1.1 and 6.1.2, the Bespalov Talanov modulation instability is studied theoretically and a minimum stripe width for instabilities to occur is derived for a focusing Kerr non-linearity. In Section 6.1.3 a coefficient for the carrier density induced Kerr-type nonlinearity is formulated under hypothetical stationary conditions. In order to investigate whether these instabilities are also obtained with the traveling wave model, simulations are performed to verify the simplified theories.

6.1.1 Bespalov Talanov modulation instability

The Bespalov Talanov modulation instability describes the breakup of a plane wave field into smaller filaments as a result of a focusing Kerr non-linearity and tiny irregularities initially present at the wave front [47]. Here, the analysis is restricted to a hypothetical steady state. The stationary states correspond to time-harmonic solutions

$$u^{\pm}(x,z,t) = u_{\mathrm{s}}^{\pm}(x,z)e^{i\Omega t} \tag{6.1}$$

of (2.24) with $\mathrm{Im}(\Omega) = 0$. For the behavior of the field within a semiconductor laser the relation between forward- u_s^+ and backward- u_s^- propagating fields has to be taken into account, whereas here the analysis is restricted to u_s^+ only, for the description of an amplifier model. Propagation in an isotropic nonlinear dielectric is considered so that only the optical Kerr nonlinearity $\Delta n_{2\mathrm{r,eff}}$ (2.37) is kept and by choosing the reference index favorably (2.24) reduces to

$$\frac{i}{2\bar{n}k_0}\partial_x^2 u_{\mathrm{s}}^{+}(x,z) = -\partial_z u_{\mathrm{s}}^{+}(x,z) - ik_0 n_2' \hbar\omega_0 v_{\mathrm{g}} d \|u_{\mathrm{s}}^{+}(x,z)\|^2 u_{\mathrm{s}}^{+}(x,z). \tag{6.2}$$

Eq. (6.2) is the starting point for the instability analysis performed in Ref. [47]. A monochromatic plane wave in the form of $u_{s0}^+(z) = u_{s00}^+ e^{-i\kappa z}$ with $\kappa = k_0 n_2' \hbar\omega_0 v_g \|u_{s00}^+\|^2$ is a solution of (6.2) when $|u_{s0}^+(z)|^2 = |u_{s00}^{+2}|$ holds. The field $u_s^+(x,z)$ is expressed as sum of this strong plane wave component u_{s0}^+ and a small perturbation δu_s^+, that represent initially present irregularities,

$$u_s^+(x,z) = (u_{s00}^+(z) + \delta u_s^+(x,z)) \cdot e^{-i\kappa z}. \tag{6.3}$$

For the small perturbations an exponential solution in the form,

$$\delta u_s^+(x,z) = u_{s1}^+ e^{-i(qx+\gamma z)} + u_{s-1}^{+*} e^{i(qx+\gamma^* z)}, \tag{6.4}$$

is assumed. By inserting (6.3) and (6.4) into (6.2), taking into account only first order terms of the perturbation and dropping fast oscillating terms $e^{\pm 2i(qx+\gamma z)}$, an expression for γ can be derived [53, 1],

$$\gamma = \pm\frac{q}{2\bar{n}k_0}\sqrt{q^2 - 4\bar{n}k_0^2 n_2' \hbar\omega v_g d \|u_{s00}^+\|^2}. \tag{6.5}$$

The system of equations will result in amplification only when $\mathrm{Im}(\gamma) > 0$. Thus irregularities in the form of Eq. (6.4) can grow only for positive real values of n_2'.

6.1.2 Instabilities induced by the optical material Kerr effect

In [47] propagation in an isotropic nonlinear dielectric is considered. In semiconductor materials n_2' can be positive or negative depending on the ratio of photon energy and energy gap [50] (*cf.* n_2 in Fig. B.2(b)). Note, that a focusing instability will only occur for $n_2' > 0$. For the investigated structures n_2' is of the order of $-1 \cdot 10^{-11}$ mW^{-1} to $-5 \cdot 10^{-13}$ mW^{-1} (with Eq. (2.37) and n_2 values from Fig. B.2(b)) and as $n_2' < 0$ here this effect is nonrelevant.

In the following a model case will still be investigated to show the mechanism in principle and to prove that instabilities of this kind are included in the time-dependent traveling model. Even by assuming pulsed operation and a high intensity of the forward propagating mode of $\hbar\omega v_g |u^+|^2 = 100$ W/µm^2 and a very high positive value of $n_2' =$

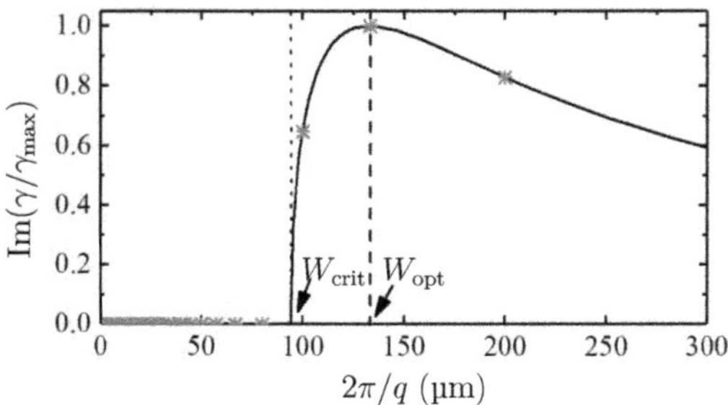

Figure 6.1: Im(γ/γ_{max}) as function of $2\pi/q$ for a field intensity of $\hbar\omega v_g\|u^+_{s00}\|^2 \approx 100$ Wμm^{-2} and $n_2' = 10^{-11}$ m/W. The vertical dotted line indicates the critical width at which instabilities occur and the vertical dashed line the optimum width for instabilities to grow. Red asterisks indicate allowed values for a total simulation domain of $W_{tot} = 400$ µm.

10^{-11} mW^{-1} results only in an index change of $\Delta n_{2r,eff} = n_2'\hbar\omega_0 v_g d(\|u\|^2 + |u^{\mp}|) = 7\cdot 10^{-6}$ with a height of the active region of $d = 7$ nm.

For this model case Fig. 6.1 shows the dependence of Im(γ/γ_{max}) on the inverse transverse wavevector $2\pi/q$. At a critical wavevector $q < q_{crit}$, with

$$q_{crit} = \sqrt{4\bar{n}k_0^2 n_2'\hbar\omega v_g d\|u^+_{s00}\|^2}, \tag{6.6}$$

the plane wave becomes unstable as Im(γ) > 0, whereas for an optimum wave vector of $q_{opt} = q_{crit}/\sqrt{2}$ a perturbation of the wave grows fastest so that under stationary conditions this perturbation would outlast all others. From q_{crit} and q_{opt} the critical

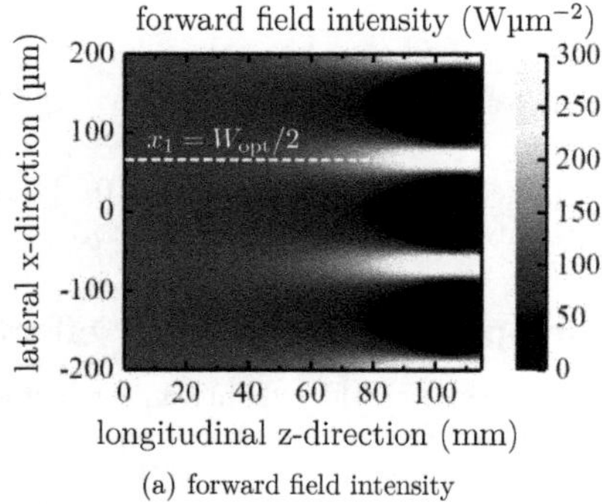

(a) forward field intensity

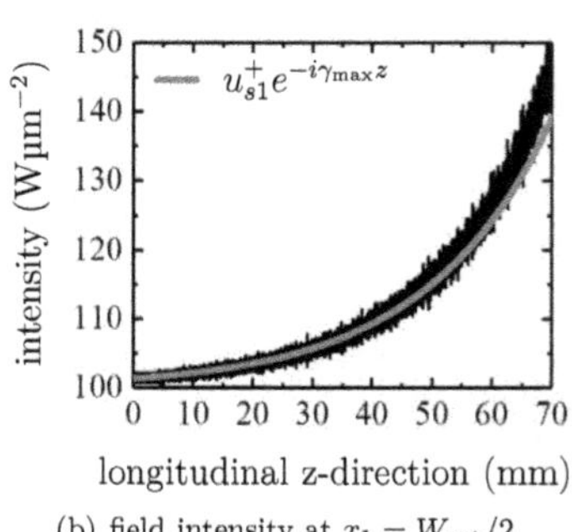

(b) field intensity at $x_1 = W_{opt}/2$

Figure 6.2: Traveling wave simulation of the forward field intensity while propagating in an isotropic nonlinear dielectric with average material Kerr coefficient $n_2' = 10^{-11}$ m/W and originally injected field intensity of $\hbar\omega v_g\|u^+_{s00}\|^2 \approx 100$ Wµm^{-2} including small random perturbations $\delta u^+_s(x, z)$. (a) Longitudinal-lateral distribution of the forward field intensity. (b) Forward propagating field intensity at position $x_1 = W_{opt}/2$ and theoretical development of $u^+_{s1}e^{-i\gamma_{max}z}$ with a filament gain of $\gamma_{max} = (21\text{ mm})^{-1}$.

transverse and optimum widths are derived as $W_{\text{crit}} = 2\pi/q_{\text{crit}} = 94$ µm and $W_{\text{opt}} = 2\pi/q_{\text{opt}} = 133$ µm, respectively (see Fig. 6.1). W_{opt} is obtained at a filament gain of $\gamma_{\text{max}} = (21 \text{ mm})^{-1}$, so that only after a comparatively long propagation distance filaments stabilizes.

In order to assess whether this theory holds for the traveling wave simulation, the following numerical experiment is performed. A high field intensity of $\hbar\omega v_g \|u^+_{s00}\|^2 \approx 100$ Wµm^{-2} including small random perturbations $\delta u^+_s(x, z = 0)$ is injected on the left side of the simulation domain and propagated for more than 100 mm through an isotropic nonlinear dielectric with average material Kerr coefficient $n'_2 = 10^{-11}$ m/W. The resulting longitudinal-lateral field intensity distribution is visible in Fig. 6.2(a). Clear filaments with a spacing that exactly corresponds to W_{opt} are observed. Due to the finite simulation domain and periodic boundary conditions only values are allowed that are a fraction of the total simulation domain $W_{\text{tot}} = 400$ µm, see red asterisks in Fig. 6.1. Furthermore Fig. 6.2(b) shows the field intensity at position $x_1 = W_{\text{opt}}/2$ together with the theoretical development of $u^+_{s1}e^{-i\gamma_{\text{max}}z}$ with a filament gain of $\gamma_{\text{max}} = (21 \text{ mm})^{-1}$. For longitudinal distances $z < 50$ mm the theoretical estimation with $\gamma_{\text{max}} = (21 \text{ mm})^{-1}$ corresponds very well to the results of the traveling-wave model. However, the more the irregularities grow, the more both curves differ due to deviations from linearity. For higher values of n'_2, W_{crit} and W_{opt} reduce.

To summarize, the traveling wave simulation reproduces very well the predicted optimum stripe width W_{opt} and filament gain γ_{max} of the simple filamentation theory. Although this proves that instabilities of this kind are included in the traveling wave model, due to the fact that $\text{Real}(n'_2) < 0$ for all structures investigated in this thesis this effect is nonrelevant. Furthermore even at the upper limit of the n'_2 coefficient and very high field intensities the effective index change is $\Delta n_{2\text{r,eff}} < 10^{-5}$. Thus it is highly doubtful that under actual laser operation with a thermally induced waveguide or built-in index variations of $> 10^{-3}$ this small value has any effect.

6.1.3 Instabilities induced by spatial hole burning

Now, instabilities arising due to an indirect Kerr-type nonlinearity in semiconductor lasers will be discussed. The Kerr-type nonlinearity arises because of spatial hole burning and the resulting depletion of carriers, which in turn results in a refractive index increase in areas of higher intensity, leading to self-focusing [2]. For this case and under hypothetical stationary conditions an equivalent Kerr coefficient will be derived as described in [27, 49].

Under hypothetical stationary conditions the carrier density can be eliminated from the equations by linearizing the real and imaginary parts of Δn^2_{eff} around a reference carrier density N_{ref}, which is typically set to the transparency carrier density N_{tr} were $g_{\text{eff}}(N_{\text{tr}}) = 0$, such that

$$g_{\text{eff}}(N) = g'_{\text{eff,lin}}(N - N_{\text{tr}}), \tag{6.7}$$

and

$$\alpha_{\text{eff}}(N) = \alpha_{\text{eff}}(N_{\text{tr}}) + \alpha'_{\text{eff,lin}}(N - N_{\text{tr}}), \tag{6.8}$$

with the linear differential gain and absorption coefficient, $g'_{\text{eff,lin}}$ and $\alpha'_{\text{eff,lin}}$, respectively. For the steady state, linearizing the recombination rate,

$$R_{\text{non-rad}} + R_{\text{spon}} = \frac{N}{\tau_N}, \tag{6.9}$$

neglecting gain dispersion, i.e. $g_r = 0$ in (2.18), as well as drift-diffusion and assuming a constant injection current density the rate equation (3.39) can be solved for the excess carrier density,

$$N - N_{\rm tr} = \frac{\frac{j\tau_N}{ed} - N_{\rm tr}}{1 + \epsilon_{\rm sat}||u_{\rm s}||^2} \tag{6.10}$$

with the saturation parameter

$$\epsilon_{\rm sat} = v_{\rm g}\, g'_{\rm eff,lin}\tau_N. \tag{6.11}$$

Inserting (6.10) into $\Delta n^2_{\rm eff,r}(N)$, $g_{\rm eff}(N)$ and $\alpha_{\rm eff}(N)$ yields

$$g_{\rm eff}(j) = \frac{g'_{\rm eff,lin}\tau_N(j - j_{\rm tr})}{ed(1 + \epsilon_{\rm sat}||u_{\rm s}||^2)}, \tag{6.12}$$

and

$$\alpha_{\rm eff}(j) = \alpha_{\rm eff}(N_{\rm tr}) + \frac{\alpha'_{\rm eff,lin}\tau_N(j - j_{\rm tr})}{ed(1 + \epsilon_{\rm sat}||u_{\rm s}||^2)}, \tag{6.13}$$

$$n^2_{\rm eff,r}(j) = n^2_{\rm eff,r}(N_{\rm tr}) + \frac{\alpha_{\rm H}\bar{n}g'_{\rm eff,lin}\tau_N(j - j_{\rm tr})}{edk_0(1 + \epsilon_{\rm sat}||u_{\rm s}||^2)} \tag{6.14}$$

with the transparency current density

$$j_{\rm tr} = edN_{\rm tr}/\tau_N \tag{6.15}$$

and Henry's α-factor

$$\alpha_{\rm H} = 2k_0\Delta n_{\rm eff,N}/g'_{\rm eff,lin}. \tag{6.16}$$

For typical values $n_{\rm g} = 4$, $g'_{\rm eff} = 10 \cdot 10^{-18}$ cm^2 and $\tau_N = 1$ ns, $\varepsilon_{\rm sat} = 7.5 \cdot 10^{-17}$ cm^3 is obtained. In general the modal gain increases with increasing carrier density whereas the effective index decreases, $g'_{\rm eff,lin} > 0$ and $\Delta n_{\rm eff,N} < 0$ for frequencies around the gain peak, and hence $\alpha_{\rm H} < 0$. Typical values of $\alpha_{\rm H}$ range from -1 to -10.

Inserting (6.1), (6.1), (6.12), (6.14) into (2.24) we obtain

$$\begin{aligned} \pm\, i\partial_z u_{\rm s}^{\pm} &= \frac{1}{2\bar{n}k_0}\partial_x^2 u_{\rm s}^{\pm}(x,z,t) + \frac{\Omega}{v_{\rm g}}u_{\rm s}^{\pm} + \frac{k_0}{2\bar{n}}\left[n^2_{\rm eff,r}(N_{\rm tr}) - \bar{n}^2\right]u_{\rm s}^{\pm} - \frac{i}{2}\alpha_{\rm eff}(N_{\rm tr})u_{\rm s}^{\pm} \\ &+ \frac{\tau_N\left[(i + \alpha_{\rm H})g'_{\rm eff,lin} - i\alpha'_{\rm eff,lin}\right](j - j_{\rm tr})}{2ed(1 + \epsilon_{\rm sat}||u_{\rm s}||^2)}u_{\rm s}^{\pm} + k_0 n'_2\hbar\omega_0 v_{\rm g} d||u_{\rm s}||^2 u_{\rm s}^{\pm}. \end{aligned} \tag{6.17}$$

For $\varepsilon_{\rm sat}||u||^2 \ll 1$ we can expand

$$(1 + \epsilon_{\rm sat}||u_{\rm s}||^2)^{-1} \approx 1 - \epsilon_{\rm sat}||u_{\rm s}||^2 \tag{6.18}$$

so that the second-last term in (6.17) resembles the last term with [49]

$$n_{2,\rm SHB} = -\frac{\tau_N\left[\alpha_{\rm H}g'_{\rm eff,lin} + i(g'_{\rm eff,lin} - \alpha'_{\rm eff,lin})\right](j - j_{\rm tr})\,\epsilon_{\rm sat}}{2ek_0\hbar\omega_0 v_{\rm g} d^2}. \tag{6.19}$$

Due to the fact that $\alpha_{\rm H} < 0$, Real($n_{2,\rm SHB}$) is positive. The focusing instabilities arise because a local intensity maximum induces a variation of the real part of the effective index with a larger value at the position of the intensity peak than outside and thus creates a local index waveguide leading to self-focusing. In a conventional laser

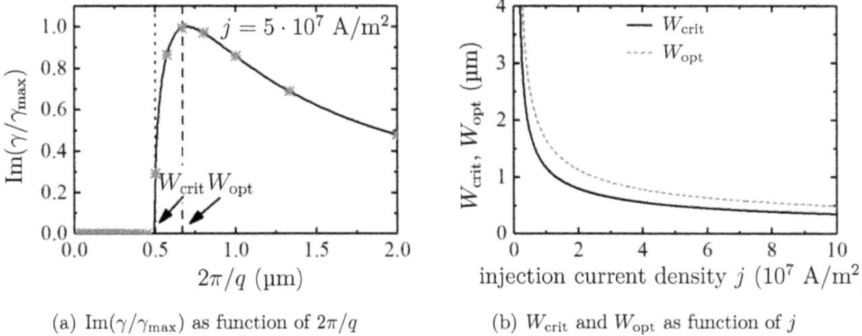

(a) Im(γ/γ_{max}) as function of $2\pi/q$

(b) W_{crit} and W_{opt} as function of j

Figure 6.3: (a) Im(γ/γ_{max}) as function of $2\pi/q$ for a field intensity of $\hbar\omega v_g \|u^+_{s00}\|^2 \approx 10$ Wμm^{-2}, $n'_2 = 0$ m/W and $n_{2,\mathrm{SHB}} = 3.6 \cdot 10^{-6}$ m/W obtained for an injection current density of $j = 5 \cdot 10^7$ A/m^2. Critical and optimum width are indicated by vertical lines. Red asterisks indicate allowed values for a total simulation domain of $W_{tot} = 4$ µm. (b) Critical W_{crit} (black solid) and optimum width W_{opt} (red dashed) as function of the injection current density j.

or amplifier medium pumped above threshold generally $g_{\mathrm{eff,lin}} > \alpha_{\mathrm{eff,lin}}$ is true and in addition to the process described here, the filament becomes destabilized because at the position of the intensity peak the modal gain is decreased which results in an reduced amplification in the local-waveguide core and an enhanced amplification outside leading to self-defocusing.

Considering structure II (Tab. A.3), linearized coefficients $\tau_N = 1$ ns and $g'_{\mathrm{eff,lin}} = 10 \cdot 10^{-22}$ m^2 and a typical injection current density for CW operation of $j = 5 \cdot 10^7$ A/m^2, $j_{\mathrm{tr}} = 0.2 \cdot 10^7$ A/m^2 (for $N_{\mathrm{tr}} = 1.7 \cdot 10^{24}$ m^{-3}) and $\alpha_{\mathrm{H}} = -1.75$ (for $N = 4 \cdot 10^{24}$ m^{-3}), Real($n_{2,\mathrm{SHB}}$) is given by Eq. (6.19) as Real($n_{2,\mathrm{SHB}}$) $= 3.6 \cdot 10^{-6}$ m/W. This is several orders of magnitude larger than the absolute value of the effective optical Kerr coefficient due to non-resonant virtual transitions which is in the range of $-1 \cdot 10^{-11}$ mW^{-1} to $-5 \cdot 10^{-13}$ mW^{-1} as pointed out above.

Thus it is not surprising that the characteristic length scales of this effect are much smaller than observed for the optical material Kerr effect as is visible in Fig. 6.3. Here the critical and optimum stripe width are $W_{\mathrm{crit}} = 0.5$ µm and $W_{\mathrm{opt}} = 0.7$ µm, respectively, for an injection current density of $j = 5 \cdot 10^7$ A/m^2 and field intensity of $\hbar\omega v_g \|u^+_{s00}\|^2 \approx 10$ Wμm^{-2} typical for CW operation. Both critical and optimum stripe width are decreasing with current, Fig. 6.3(b). Accordingly for $j = j_{\mathrm{th}} = 4 \cdot 10^6$ A/m^2 and a corresponding intensity of 1.2 Wμm^{-2} for example the critical stripe width is increased to 10 µm, which lies in the same range as calculated by [86] were a more sophisticated stability analysis is performed.

The numerical experiment performed in the former section is now repeated for a plane wave injected at the rear facet with an intensity of 10 Wµm^{-2} including small random perturbations $\delta u^+_s(x, z)$ and an injected current density of $j = 5 \cdot 10^7$ A/m^2. In this case the model parameters are adjusted so that $\partial_N g(\bar{N}) = \partial_N \alpha(\bar{N})$ and $g(\bar{N}) = \alpha(\bar{N})$ for the averaged carrier density of $\bar{N} = 3 \cdot 10^{24}$ m^{-3} to minimize the impact of modal gain changes on the field stability. Note, that in the traveling wave model a logarithmic gain model (2.31) is employed, so that this condition only holds for $\bar{N}$. If instabilities arise

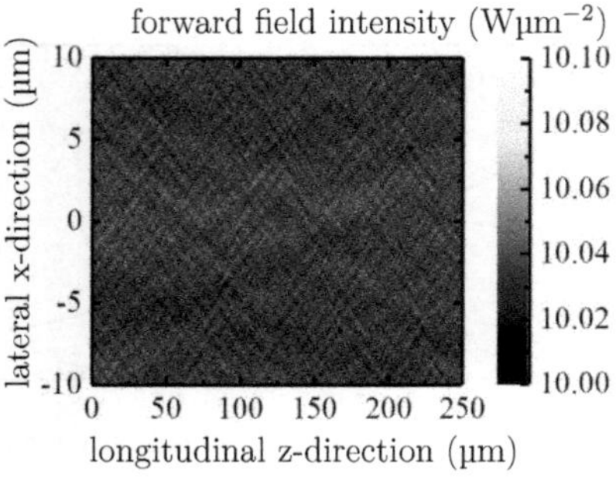

Figure 6.4: Traveling wave simulation of the forward field intensity propagating though a semiconductor gain material with an averaged carrier density of $\bar{N} = 3 \cdot 10^{24}$ m^{-3} for which the modal gain is compensated for by the absorption ($g(\bar{N}) = \alpha(\bar{N})$). An originally injected field intensity of $\hbar\omega v_g \|u^+_{s00}\|^2 \approx 10$ Wμm^{-2} including small random perturbations $\delta u^+_s(x, z)$ is assumed.

and result in a fluctuation of the carrier density, these conditions would not be met anymore, as is the case for actual laser operation. The simulation domain was chosen to have a width of 20 μm and length of 250 μm. Stabilized filaments are expected to occur after longitudinal propagation of 0.5 μm, which is very well covered by the simulation region.

After a simulation time of 500 ps stable filaments are not visible, Fig. 6.4. This indicates that due to the additional defocusing induced by changes of the modal gain and the non-stationarity of the model, stable stationary filaments can't stabilize even under such beneficial conditions as considered here.

With respect to actual laser operation it should be taken into account that the derived equations are only formulated for the forward propagating field under hypothetical stationary conditions. However, in BA lasers the highly dynamic interaction of forward- and backward-propagating fields have to be taken into account. Furthermore the assumption of a constant injection current density, i.e. an infinite series resistance as discussed in Section 3.2.2, and the neglect of drift-diffusion result in an overestimation of spatial hole burning.

The results obtained here coincide also with the single shot streak camera measurements shown in Fig. 5.4(a), where no stable filaments are visible. The observed non-stationary behavior can be better understood in the mode picture as discussed in the next section. Besides carrier density induced refractive index changes the complex effective index also depends on the temperature Δn_T and additionally built-in index variations. These waveguides stabilize the linear guided modes, which are thus observable even far above threshold [90].

6.1.4 Conclusions

A simple theory for filamentation as a result of a focusing Kerr non-linearity was revisited as formulated in [47]. It was shown that this mechanism is included in the traveling-wave equations and a numerical experiment could support the simplified theory. For the utilized wavelength range and material compositions, however, the value of the direct optical Kerr nonlinearity is small and not leading to focusing, rendering it irrelevant.

For semiconductor lasers an equation for an indirect Kerr-type coefficient $n_{2,\mathrm{SHB}}$ can be formulated for a hypothetical steady state. The Kerr-type non-linearity results from carrier density reduction in areas where the field intensity is high (spatial hole burning), because the refractive index in these areas rises, thus creating a local waveguide. Compared to the absolute value of the direct optical Kerr nonlinearity discussed before, $\mathrm{Re}(n_{2,\mathrm{SHB}})$ is several orders of magnitudes larger and theoretically leads to focusing.

In the simple theory presented here, changes of the modal gain are neglected, which have a defocusing effect, because the modal gain decreases in areas of high field intensity resulting in reduced amplification. Furthermore, the theory was formulated under hypothetical stationary conditions for the forward propagating field alone. In BA lasers the interaction of forward- and backward-propagating fields and their non-stationarity have to be considered. Consequently, a simulation of the forward field intensity propagating though a semiconductor gain material shows no formation of stable filaments, which also coincides with experimental results. The non-stationary behavior of BA lasers can possibly be better understood in the mode picture as discussed in the next section.

6.2 Multi-mode lasing

The phenomena occurring in BA lasers can be understood in the mode picture [88, 92, 20]. In index-guided lasers lateral modes can be identified by spectrally-resolved near and far-field measurements [89, 16]. In the traveling-wave approach (2.24) no pre-assumptions are made regarding lasing modes. However, any field can be expressed as a linear combination of a complete set of eigenmodes. To gain a clearer understanding of the field obtained by the traveling-wave approach, in a post-processing the optical field can be decomposed into lateral waveguide modes. The procedure is exemplarily shown in this section.

The lateral waveguide modes satisfy the Helmholtz equation for a hypothetical steady state,

$$\left[k_0^{-2}\partial^2/\partial x^2 + \langle n_{\mathrm{eff}}(x,z_0)\rangle_t^2\right]\phi_m(x,z_0) = \hat{n}_m^2\phi_m(x,z_0), \tag{6.20}$$

where z_0 is a fixed longitudinal position, $\hat{n}_m$ is the modal index being the complex-valued eigenvalue and $\langle n_{\mathrm{eff}}(x,z_0)\rangle_t = \langle \bar{n} + \Delta n_{\mathrm{eff}}(x,z_0)\rangle_t$ with Δn_{eff} from Eq. (2.25) the time averaged complex valued effective index, which is gained in post-processing from the time averaged profiles of carrier density, temperature and field intensity.

The complex forward- or backward-propagating fields $u^\pm$ at position z_0 and t_0 can now be expressed as a linear combination (lc) of the eigenfunctions $\phi_m(x,t_0,z_0)$ of Eq. (6.20),

$$u_{\mathrm{lc}}^\pm(x,t_0,z_0) = \sum_m a_m^\pm(t_0,z_0)\phi_m(x,t_0,z_0) \tag{6.21}$$

with the modal coefficients

$$a_m^\pm(t_0,z_0) = \frac{\int \phi_m(x,t_0,z_0)u^\pm(x,t_0,z_0)\,dx}{\int \phi_m^2(x,t_0,z_0)\,dx}. \tag{6.22}$$

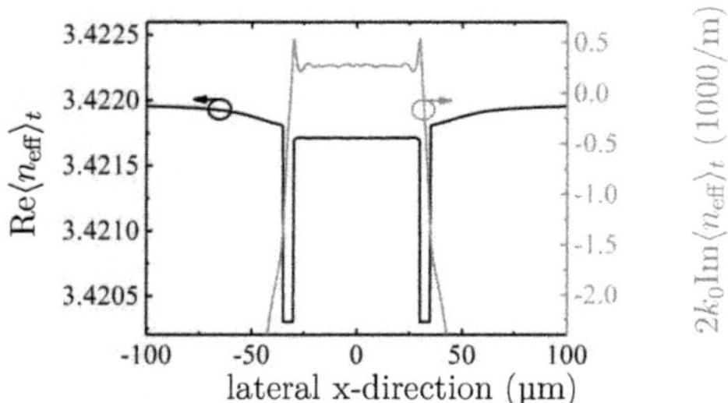

Figure 6.5: Time averaged profiles of the real part of the effective index $\mathrm{Re}\langle n_{\mathrm{eff}}(x)\rangle_t$ (left black axis) and effective gain $2k_0\mathrm{Im}\langle n_{\mathrm{eff}}(x)\rangle_t$ (right red axis).

To assess whether the linear combination of a certain number of modes is a good approximation to the field, the following error is defined,

$$\sigma^{\pm}(t_0, z_0) = \frac{\int |u^{\pm}_{\mathrm{lc}}(x, t_0, z_0) - u^{\pm}(x, t_0, z_0)|^2 dx}{\int |u^{\pm}(x, t_0, z_0)|^2 dx}. \tag{6.23}$$

In the following the decomposition is shown exemplarily for a laser with an injection stripe width of $W = 50$ µm and length $L = 4$ mm (structure III Tab. A.6) and injection current of $I = 24.1$ A resulting in a total output power of 25.6 W under isothermal conditions. The structure is considered to be unimplanted and index-guiding trenches are etched at $|x| \in [30 : 35]$ µm ($\Delta n_0 = -1.7 \cdot 10^{-3}$). The total simulation time is 8 ns, whereas over the last 5 ns of this time period the averaged quantities are obtained.

The time averaged profiles of the real part of the effective index and gain are shown in Fig. 6.5. The imaginary and real parts of the eigenvalues of (6.20) are displayed in Fig. 6.6 where the left axis is the modal gain $g_m = 2k_0\mathrm{Im}(\hat{n}_m)$ and the abscissa the real part of the modal index relative to the real part of the index of the fundamental mode, $\Delta\mathrm{Re}(\hat{n}_m) = \mathrm{Re}(\hat{n}_m) - \mathrm{Re}(\hat{n}_1)$ (black dots). The upper "arm" with high modal gain corresponds to guided modes that are confined to the region below the injection stripe. On the right axis the time averaged relative modal coefficients $|a_m|^2/\sum_m |a_m|^2$ of the

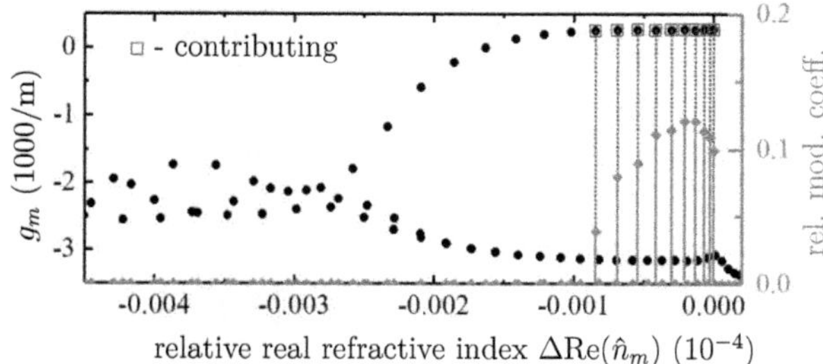

Figure 6.6: Real (abscissa) and imaginary (ordinate) eigenvalues of the modes obtained from the solution of the Helmholtz equation (6.20). Eigenvalues of modes with a significant contribution to the field (*cf.* Fig. 6.7(c)) are marked blue.

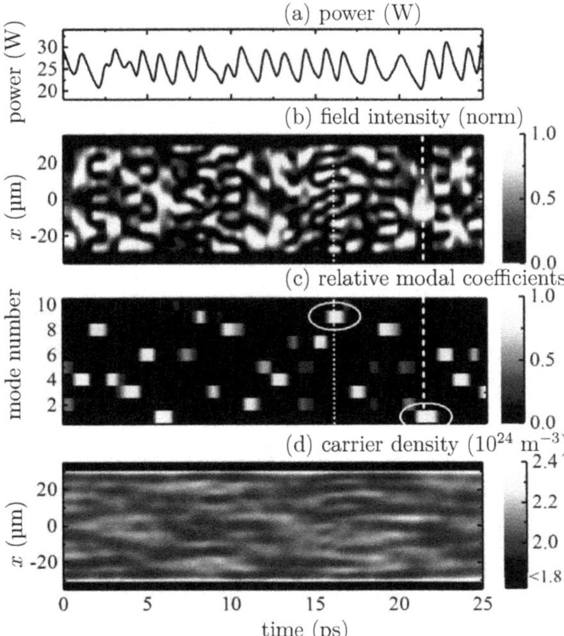

Figure 6.7: (a) Temporal evolution of the optical output power, (b) pseudo-color mapping of the near-field intensity distribution (normalized to maximum in every time step) and (c) similar mapping of the relative modal coefficients and (d) carrier density distribution at the front facet. The white vertical lines indicate stages where either a high order mode (dotted) or the fundamental mode (dashed) dominates.

modes are displayed (red diamonds). All modes that have a significant contribution to the field (*cf.* Fig. 6.7(c)) belong to those upper "arm" modes and are marked blue. Those modes are needed for the reproduction of the field with a time averaged error (6.23) of $\langle \sigma^+(t, z_0) \rangle_t < 1\%$, where $\langle \rangle_t$ denotes the time-average. The lower "arm" corresponds to (semi-)radiation modes, that are not confined to the pumped laser region and oscillate in the regions beside the injection stripe. In an open geometry those modes are a continuous set.

Fig. 6.7 shows the result of the modal decomposition of the forward propagating field at the output facet $u^+(x, t, z_0 = L)$ for a short time period of 25 ns. Panel (c) displays the magnitudes of the relative modal coefficients $|a_m|^2 / \sum_m |a_m|^2$ versus time as a pseudo-color mapping for the 10 modes with the highest gain. Those are the only modes that have a significant contribution to the field (marked blue in Fig. 6.6). It is visible that the strong dynamic behavior of the emitted power and near-field intensity (see Fig. 6.7(a) and 6.7(b)) can be explained by alternate lasing of different lateral modes. It is notable that in this example the different modes actually lase alternately and comparatively few modes lase at the same time. At most time instances the lasing mode can directly be traced back to the dynamic lateral structure of the intensity. For example, at $t = 16.2$ ps (white dotted), the near-field in panel (b) has 9 maxima and

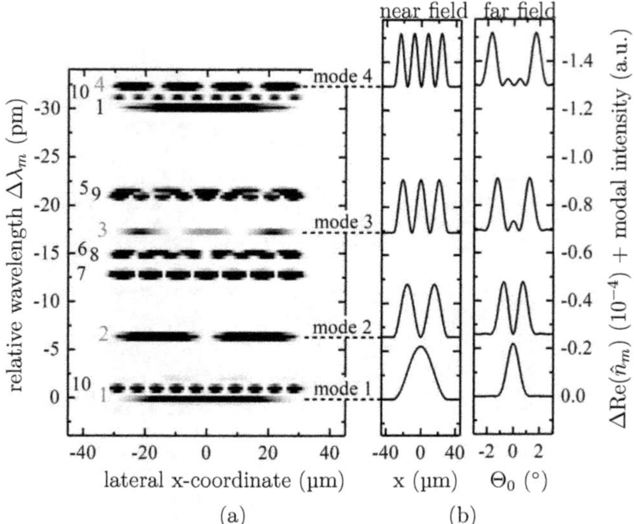

Figure 6.8: (a) Intensity distribution of the spectrally resolved complex near-field obtained by the traveling wave model as function of the wavelength relative to the peak wavelength of the fundamental mode. On the left side the lateral mode number is displayed. (b) near- and far-field intensities of the first four lateral modes obtained by solving Eq. (6.20) vertically shifted according to their real part of the modal index relative to the real part of the index of the fundamental mode.

accordingly the 9th mode has the highest contribution to the field, whereas at $t = 21.4$ ps (white dashed) the near-field has one maximum and accordingly the fundamental mode has the highest contribution.

This strongly contrasts the findings of [93], where the lateral carrier density variation reaches values of up to $\Delta N = 2 \cdot 10^{24}$ m^{-3} near the front facet at moderate injection currents. The results obtained in [93] were calculated with a wave optical beam propagation algorithm, disregarding the non-stationarity of the optical field in BA lasers and might in turn lead to an overestimation of the lateral carrier density variation.

In Fig. 6.8(a) the spectrally resolved near-field intensity profiles obtained from the traveling wave model by a Fourier transformation (FT) of $u^+(x, t, z_0 = L)$ are shown. The lateral mode number is given on the left side. The longitudinal mode spacing of $\Delta\lambda_{\text{FT}}^{\text{long}} = 30$ pm is larger than the spacing of low order lateral modes and not all lateral modes displayed in Fig. 6.8(a) correspond to the same longitudinal mode.

The lateral intensity profiles of the near and far field of the first four modes with the highest gain according to Fig. 6.6 obtained from the solution of (6.20) are exemplary displayed in Fig. 6.8(b). The zero levels of the mode intensities are given by the relative real parts of the corresponding modal indices which can be converted into corresponding relative wavelengths $\Delta\lambda_m = \lambda_m - \lambda_1 = \lambda_0/n_{\text{g}} \cdot \mathrm{Re}(\Delta\hat{n}_m)$ where $n_{\text{g}} = 3.92$ is the group index. As indicated by the dotted lines, these wavelength shifts agree very well with the FT peak wavelengths of the lateral modes for the same longitudinal mode (colored red in Fig. 6.8(a)). Thus, although no pre-assumptions were made regarding the optical field the results indicate that in lasers with a lateral waveguide a clear mode structure is visible which is neither destroyed by the dynamics nor by longitudinal effects.

Close to laser threshold the wavelength mode spacing relative to the fundamental mode wavelength λ_0 can be approximated theoretically by assuming abrupt perfectly reflecting walls [94, 16]. For the vertical mode order 1, longitudinal mode order l and lateral mode order m it is given by [95, 94]

$$\Delta\lambda_{\text{theory}} = -\frac{\lambda_0}{2n_g}\left(\frac{l}{L}\lambda_0 + \frac{(m^2-1)\lambda_0^2}{4n_{\text{eff}}W^2}\right), \tag{6.24}$$

so that the equidistant longitudinal mode spacing for the same lateral mode is $\Delta\lambda_{\text{theory}}^{\text{long}} = \lambda_0^2/(2Ln_{\text{g}})$ and the lateral mode spacing for modes m and $m{+}1$ and the same longitudinal mode is given by $\Delta\lambda_{\text{theory}}^{\text{lat } m/m+1} = (2m+1)\lambda_0^3/(8n_{\text{eff}}n_gW^2)$. In the example at hand the longitudinal mode spacing according to (6.24) is predicted to be $\Delta\lambda_{\text{FT}}^{\text{long}} = 30$ pm, which exactly corresponds to the longitudinal modes spacing obtained by the traveling-wave model. The lateral mode spacing from the second to the fundamental mode, $\Delta\lambda_{\text{theory}}^{\text{lat } 1/2} = 7$ pm with the aperture $W = 60$ µm and effective refractive index in the waveguide of $n_{\text{eff}} \approx 3.4$, also agrees well with the mode spacing of $\Delta\lambda_2 = 6.4$ pm in Fig. 6.8(a). However, the theoretical mode spacing deviates more and more from those values of Fig. 6.8(a), which can be expected as Eq. (6.24) is deducted making very simplified assumptions for the lateral resonator. During the discussion of Fig. 6.8(a) it was already pointed out that the longitudinal mode spacing is larger than the spacing of low order lateral modes. From Eq. (6.24) it is now visible that this is an attribute of BA lasers, whereas ridge waveguide lasers with W in the range of some micrometers have a larger lateral than longitudinal mode spacing.

In conclusion, in the traveling-wave approach no pre-assumptions regarding modes is made, however, any field can be expressed as a linear combination of a complete set of eigenmodes. Here in a post-processing the optical field obtained by a laser simulation with the traveling-wave model is decomposed into lateral waveguide modes. The results indicate that in lasers with a lateral waveguide a clear mode structure is visible which is neither destroyed by the dynamics nor by longitudinal effects.

6.3 Nonthermal effects

In the absence of a thermally induced waveguide, transverse instabilities arise due to lateral spatial hole burning and the nonlinear interaction between the electromagnetic field and the gain material. The controlling parameters are those influencing the carrier reservoir, such as series resistivity (3.46), sheet resistance (3.47), and mobility, as well as the interaction between electromagnetic field and gain, the differential index n'_N and gain g'.

With increasing current, instabilities become more pronounced and the far field broadens, which is often termed far-field blooming. Non-thermal far-field blooming becomes the dominant effect under pulsed operation with current pulses long compared to the turn-on time combined with a very small repetition rate to avoid device heating. In the mode picture this can be explained by an increased number of excited lateral modes. This thesis has been supported by a stationary model based on the incoherent superposition of multiple modes [21]. However, BA lasers exhibit an inherently non-stationary behavior and the assumption of stationary multimode conditions is

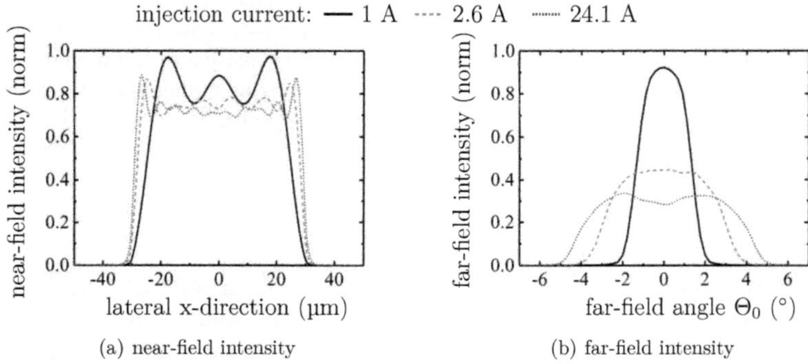

(a) near-field intensity (b) far-field intensity

Figure 6.9: Time averaged normed lateral (a) near and (b) far-field distribution at the front facet for different injection currents. The fields are normed, i.e. divided by their lateral integral. $r_s = 5 \cdot 10^{-9}$ Ωm^2 and $d_{\text{res,imp}} = 1345$ nm.

questionable [54]. In the traveling-wave approach no pre-assumptions regarding lateral modes is made.

In this section the effect of non-thermal far-field blooming and the dependencies of transverse instabilities regarding the previously discussed control parameters is investigated. The investigated structure is the same as in Section 6.2 with an injection stripe width of $W = 50$ µm and length $L = 4$ mm (structure III Tab. A.6) and index-guiding trenches etched at $|x| \in [30 : 35]$ µm ($\Delta n_0 = -1.7 \cdot 10^{-3}$).

6.3.1 Nonthermal far-field blooming

The effect of non-thermal far-field blooming is exemplary investigated for a laser with the same parameters as are used in Section 6.2. The total simulation time is $\tau_{\text{sim}} = 8$ ns and over the last $\tau_{\text{av}} = 5$ ns of this time period the averaged quantities are obtained.

In Fig. 6.9 the time averaged near- and far-field intensity distributions at the front facet are exemplarily displayed for the total injection currents $I = 1$ A, 2.6 A and 24.1 A.

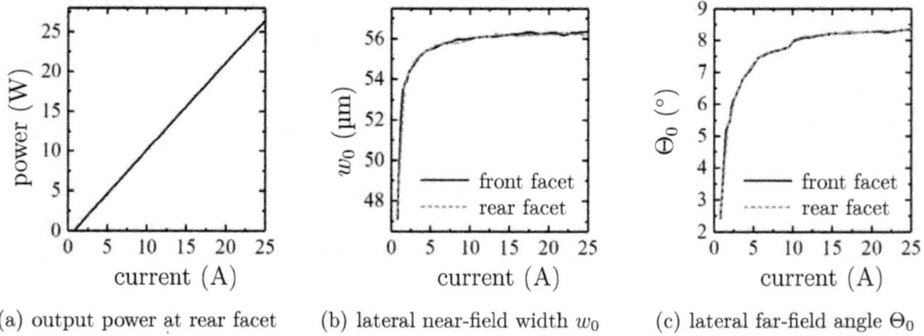

(a) output power at rear facet (b) lateral near-field width w_0 (c) lateral far-field angle Θ_0

Figure 6.10: Output power at rear facet (a) and lateral near-field width w_0 (b) and far-field angle Θ_0 (95 % power content) (c) at front (black solid) and rear (red dashed) facet.

It is visible that near threshold at $I = 1$ A the near field is not single-lobed. The fields are normed, i.e. divided by their lateral integral, and thus it is easily visible that the near and far fields widen with current.

To investigate this broadening in more detail the full near-field width w_0 and far-field angle Θ_0 containing 95% of the power are extracted from the time averaged near and far-field intensities at the front and rear facet. Fig. 6.10 shows their dependence on the injection current for the front and rear facet, respectively. Both near-field width and far-field angle rise with increasing injection current. However, the highest increase is visible within the first 5 A, after that the widths remain approximately unchanged. At approximately 10 A a small step increase is visible in the far-field angle. Interestingly at both facets the field widths show the exact same behavior. Later, in Section 6.4.2 it becomes clear that this is uniquely the case for strongly-index-guided laser operation.

To gain a better understanding of the increased near-field width and far-field angle, the modal decomposition of the field at the front facet, as described in Section 6.2, is performed. As stated there at the highest displayed current of $I = 24.1$ A laser emission is made up by the 10 modes with the highest gain. The relative modal coefficients $|a_m|^2/\sum_m |a_m|^2$ for these modes are displayed in Fig. 6.11 as function of the current. Directly above threshold ($I_{\text{th}} = 0.7$ A) laser emission is made up mostly by the first mode. With increasing current its relative contribution is diminished (without turning off completely), as more and more modes reach threshold. This behavior differs from what was published earlier [21] where at threshold the first 3 modes have an equal contribution to lasing and then consecutively turn off as more modes emerge. The rapid increase of contributing modes is taking place within the first 5 A where the first 9 modes are excited, so that the far-field angle is significantly broadened during this current span. A further abrupt rise of the far-field angle between 9.5 A and 10 A (Fig. 6.10) can be traced back to the excitation of the 10th mode.

Thus, non-thermal far-field blooming for strongly index-guided lasers can be understood on the basis of superposition of stationary modes despite the highly non-stationary behavior of BA lasers.

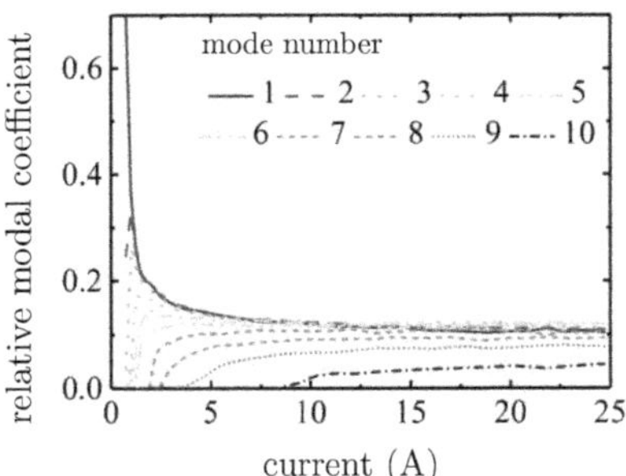

Figure 6.11: Time averaged mean square values of the relative modal coefficients $|a_m|^2/\sum_m |a_m|^2$ as function of the injection current for the field decomposition at the front facet.

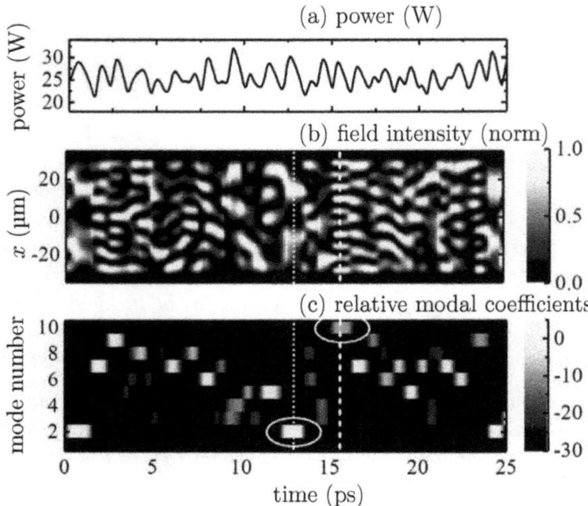

Figure 6.12: Same as Fig. 6.7 but with $n'_N = 0$ m^3. (a) Temporal evolution of the optical output power, (b) pseudo-color mapping of the near-field intensity distribution (normalized to maximum in every time step) and (c) similar mapping of the relative modal coefficients. The white vertical lines indicate stages where either a high order mode (dashed) or the fundamental mode (dotted) dominates.

6.3.2 Differential index (α_{H}-factor)

The interaction between electromagnetic field and gain, in this model quantified by the differential index n'_N (2.29) and gain g' (2.31), is often described by the Henry α_{H}-factor, which in our model framework is given by $\alpha_{\text{H}}(N) = \partial_N \text{Re}(n_{\text{eff}}) \big/ \partial_N \text{Im}(n_{\text{eff}}) = -k_0\sqrt{n'_N N}/g'$. As the α_{H}-factor is carrier dependent, in this section the focus will be on the influence of the differential index n'_N on the field distribution inside the resonator.

As discussed in Section 6.1 the carrier density induced effective Kerr coefficient (6.19) is several orders of magnitude larger than the absolute value of the material Kerr coefficient [50], so that the latter one can be ignored. Thus in this analysis $\Delta n'_2 = 0$ will be set, so that the focus lies on the effect of spatial hole burning and its interaction with the optical field.

In the following, the influence of a vanishing α_{H}-factor is investigated, i.e. $n'_N = 0$. This means that the Kerr-type non-linearity, which is induced by spatial hole burning and also discussed in Section 6.1.3, is neglected, $\text{Re}(\Delta n_{2,\text{SHB}}) = 0$. The laser is turned on by a 5 ns long voltage ramp and then simulated for another $\tau_{\text{sim}} = 10$ ns. During the last $\tau_{\text{av}} = 5$ ns the time averaged data is obtained, whereas during the last 1 ns the laser time trace displayed in Fig. 6.12 is obtained. In Fig. 6.12 no qualitative difference is visible compared to Fig. 6.7. Similar to the prior simulation 10 modes reach threshold and contribute to the field by alternate lasing, Fig. (6.12)(c). At most time instances the field intensity can be easily traced back to a certain lateral mode. The mode interaction leads to a high dynamic and multi-peaked field intensity. Thus the carrier induced Kerr type nonlinearity is at least not alone responsible for the multi-peaked

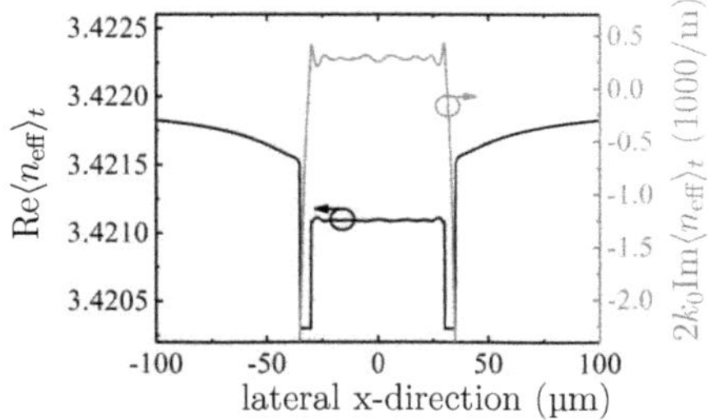

Figure 6.13: Same as Fig. 6.5 but with $n'_N = 40 \cdot 10^{-32}$ m^3. Time averaged profiles of the real part of the effective index $\mathrm{Re}\langle n_{\mathrm{eff}}(x)\rangle_t$ (left black axis) and gain $2k_0\mathrm{Im}\langle n_{\mathrm{eff}}(x)\rangle_t$ (right red axis).

field intensity, which agrees with the findings of Section 6.1.3, where the simplified theory of filamentation is found to have a minor impact on the field formation.

In a further study the differential index is now increased by one order of magnitude to $n'_N = 40 \cdot 10^{-32}$ m^3. The resulting time averaged profiles of the real part of the effective index and gain are shown in Fig 6.13. Due to the higher n'_N a slight modulation of the real part of the effective index is visible in the stripe region compared to Fig. 6.5. However, as discussed earlier with regards to Fig. 6.7(d) the lateral carrier density variation below the injection stripe is small compared to the overall difference from injection stripe to the unpumped regions. As a consequence, the overall index below the stripe is significantly decreased and the slight modulation of the real part of the effective index below the stripe is negligible.

Decomposing the field at the front facet into lateral modes of the time-averaged waveguide was performed as described in Section 6.2. The modal gain and time average of the relative modal coefficients are displayed in Fig. 6.14. Here, also lower gain modes are needed to reproduce the field with an error (6.23) $\langle\sigma^+(t, z_0)\rangle_t < 1\%$, marked blue in Fig. 6.14. Due to anti-index guiding all guided modes extend to a larger amount into

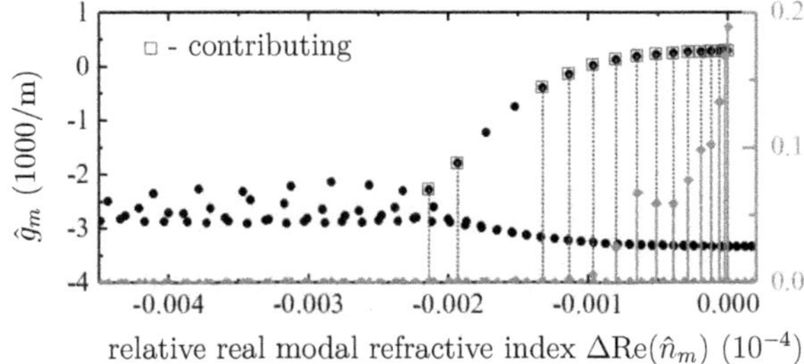

Figure 6.14: Same as Fig. 6.6 but with $n'_N = 40 \cdot 10^{-32}$ m^3. Real (abscissa) and imaginary (ordinate) eigenvalues of the modes obtained from the solution of the Helmholtz equation (6.20). Eigenvalues of modes with a significant contribution to the field (*cf.* Fig. 6.15(c)) are marked blue.

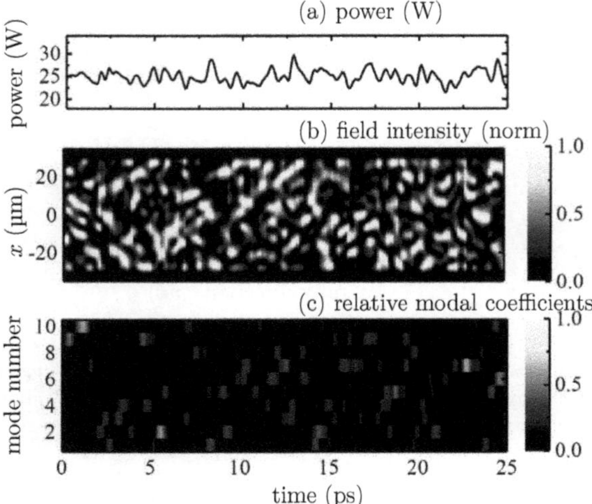

Figure 6.15: Same as Fig. 6.7 but with $n'_N = 40 \cdot 10^{-32}$ m^3. (a) Temporal evolution of the optical output power, (b) pseudo-color mapping of the near-field intensity distribution (normalized to maximum in every time step) and (c) similar mapping of the relative modal coefficients.

the unpumped regions, so that higher order modes have higher losses and the relative modal coefficients of low order modes rise compared to Fig. 6.6.

The relative modal coefficients $|a_m|^2/\sum_m |a_m|^2$ for the 10 modes with the highest gain are displayed in Fig. 6.15(c) together with the dynamics field evolution 6.15(b). Compared to the similar mapping in Figs. 6.7 and 6.12 the field is now described by the alternate lasing of modes on a much smaller time scale and the spatial and temporal scales of the intensity structures become smaller. By further increasing n'_N the spatial and temporal scales of the intensity structures become even smaller, Fig. 6.16.

In Fig. 6.17 the time-averaged near and far fields are displaying for varying differential index values n'_N for comparison. Although the lateral near-field structure might suggest a higher degree of "filamentation“ with increased differential index, the occurrence of distinguishable stable filaments is also not observed at an extremely high differential index of $n'_N = 200 \cdot 10^{-32}$ m^3, Fig. 6.16, using the traveling-wave model. Note that for $n'_N = 200 \cdot 10^{-32}$ m^3 the laser becomes anti-index guided (i.e. the effective index inside

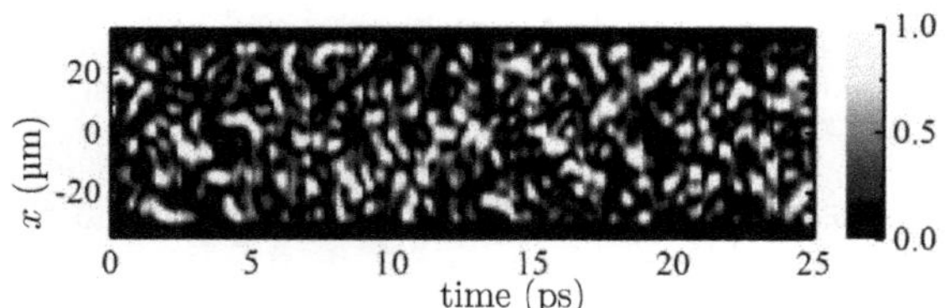

Figure 6.16: Same as Fig. 6.7(c) but with $n'_N = 200 \cdot 10^{-32}$ m^3. Pseudo-color mapping of the near-field intensity distribution (normalized to maximum in every time step).

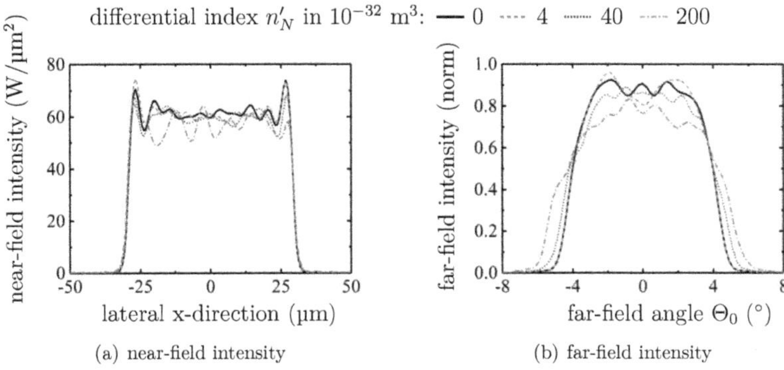

(a) near-field intensity

(b) far-field intensity

Figure 6.17: Time averaged lateral (a) near and (b) far-field distribution at the front facet for different injection currents. The far fields are divided by their lateral integral. $r_s = 5 \cdot 10^{-9}$ Ωm^2 and $d_{\text{res,imp}} = 1345$ nm.

the contact stripe is smaller than within the index-guiding trench). Consequently, the modal structure of the output field becomes more complicated and the far field slightly broadens, Fig. 6.17(b).

In conclusion, by neglecting the Kerr-type non-linearity (i.e. setting $n'_N = 0$) the emitted field is still multi-peaked and shows a high dynamic, suggesting that the filamentation mechanism described in [47] is at least not alone responsible for the "filamented" near-field structure in BA lasers. By increasing the differential index n'_N to $4 \cdot 10^{-31}$ more modes contribute to the field and the spatial and temporal scales of the intensity structures become smaller. However, the occurrence of distinguishable stable filaments is also not observed at an extremely high differential index of $n'_N = 2 \cdot 10^{30}$ m^3, Fig. 6.16, using the traveling-wave model.

6.3.3 Lateral carrier distribution in the active region

In this subsection, the focus lies on the effect the carrier reservoir has on transverse instabilities and thus the beam quality. The governing parameters that are investigated with respect to far-field blooming are the series resistivity (3.46), sheet resistance (3.47) and mobility. The investigated structure is the same as in the previous sections with an injection stripe width $W = 50$ µm and length $L = 4$ mm (structure III Tab. A.6) and index-guiding trenches etched at $|x| \in [30 : 35]$ µm ($\Delta n_0 = -1.7 \cdot 10^{-3}$).

In order to suppress higher order lateral modes and to improve the beam quality of the laser an additional implantation beside the contact stripe [22] can be considered, that increases the sheet resistance (3.47). As shown in Fig. 6.18(a), already an implantation of the contact layer drastically improves the beam quality, as the number of lateral lasing modes is decreased from 10 to 4 (Fig. 6.18(b)). In the absence of thermal lensing a further implantation of cladding and confinement layers leads only to minor improvements rendering deep implantation unnecessary for pulsed operation (if index-guiding trenches close to the stripe are present). Note that the residual layer thicknesses of 5, 770, 1215 and 1345 nm result in currents of $I = 20.0$, 20.4, 21.5 and 24.1 A and powers of $P = 21.3$, 21.6, 22.8 and 25.6 W, respectively, for a constant $U = 1.89$ V.

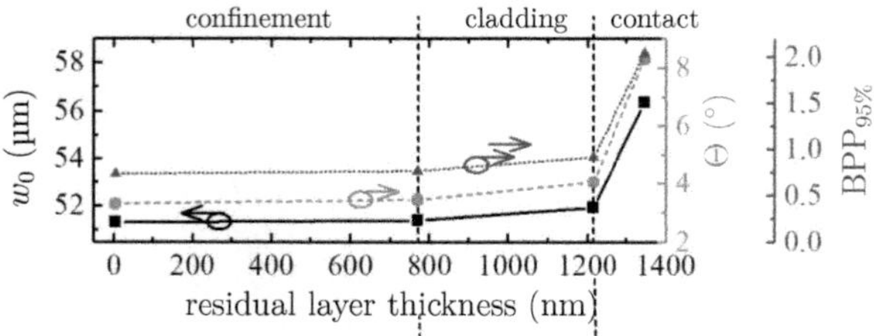

(a) lateral near-field width (w_0), far-field angle (Θ) and beam parameter product (BPP) as function of residual layer thickness $d_{\mathrm{res,imp}}$

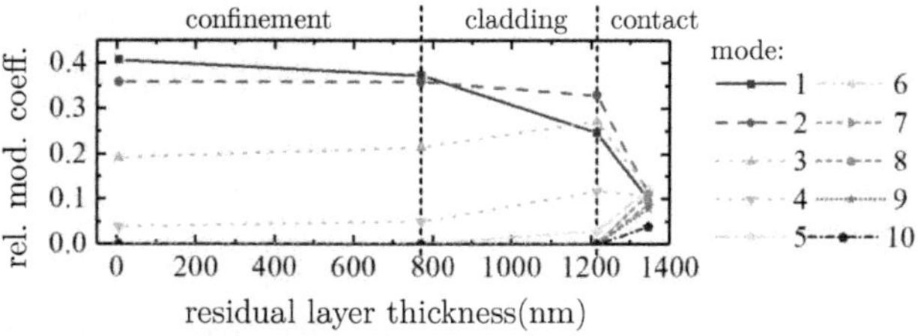

(b) relative modal coefficients as function of residual layer thickness $d_{\mathrm{res,imp}}$

Figure 6.18: Impact of decreased residual layer thickness on the beam quality and mode composition. (a) Lateral near-field width w_0 (black line - left axis), far-field angle Θ_0 (red dashed - right axis) and corresponding beam parameter product BPP (blue dotted - rightmost axis) and (b) relative modal coefficients as function of the residual layer thickness $d_{\mathrm{res,imp}}$ for $U = 1.89$ V. Series resistivity $r_s = 5 \cdot 10^{-9}$ $\Omega\mathrm{m}^2$.

Due to the fact that the excitation of higher order modes is caused by the saturation of the gain in those parts of the active layer where the lower order mode intensity is high and a rise of the gain in the other parts, a suppression of spatial hole burning should result in an improved beam quality. In the study of [3] it is demonstrated that a reduction of the p-layer resistivity r_s results not only in an equalized carrier density below the contact stripe due to current self-distribution [64, 33] but can also lead to an improvement of the beam quality. Fig. 6.19(a) reveals that the near-field width w_0 and far-field angle Θ_0 are decreased by reducing r_s. A modal analysis confirms that 4 lateral modes are lasing at $r_s = 5 \cdot 10^{-9}$ $\Omega\mathrm{m}^2$ and the fundamental mode has a contribution of less than 50 %, Fig. 6.19(b). However, at $r_s = 5 \cdot 10^{-11}$ $\Omega\mathrm{m}^2$ 90 % of laser emission is made up by the first and 10 % by the second mode, so that the theoretical limit of the beam parameter product of BPP$= \lambda_0/\pi = 0.31$ mm mrad is nearly reached.

Theoretically spatial hole burning and the lateral spreading of the carrier density is also influenced by changes in the conductivity of carriers in the active region (3.35) and thus the effective diffusion coefficient (3.33). The influence of the varying mobility entering Eq. (3.35) on the time-averaged distribution of the carrier density can be seen in Fig. 6.20.

Two counteracting effects have to be considered when discussing the influence of the hole mobility on the widths of near and far-field intensity: On the one hand, a higher mobility and thus diffusion coefficient should increase near-field width and far-field angle as the carriers spread further into the unpumped regions, where they can recombine. On the other hand, it also leads to reduced spatial hole burning because fluctuations

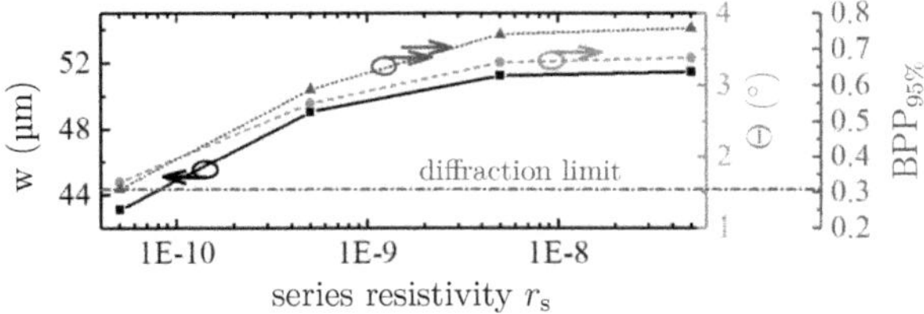

(a) lateral near-field width (w_0), far-field angle (Θ) and beam parameter product (BPP) as function of residual layer thickness $d_{res,imp}$

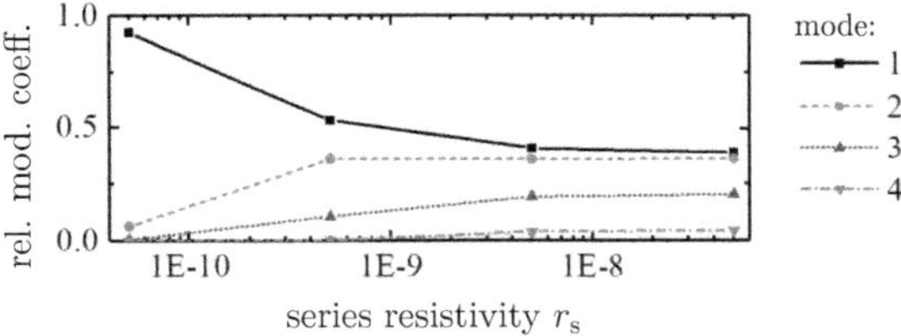

(b) relative modal coefficients as function of residual layer thickness $d_{res,imp}$

Figure 6.19: Impact of decreased series resistivity on the beam quality and mode composition. (a) Lateral near-field width w_0 (black line - left axis), far-field angle Θ_0 (red dashed - right axis) and beam parameter product BPP (blue dotted - rightmost axis) and (b) relative modal coefficients as function of the series resistivity r_s for an injection current of $I = 20.0$ A corresponding to an output power of $P \approx 21.3$ W. Residual layer thickness $d_{res,imp} = 5$ nm.

in the carrier density can be faster compensated for, as is visible in the inset of Fig. 6.20, and should thus be beneficial for the beam quality. From Fig. 6.21 it is clear that the latter effect theoretically predominates. However, increasing or reducing the value of the hole mobility of $\mu_p \approx 0.03$ m^{2}V^{-1}s^{-1} by one order of magnitude has no

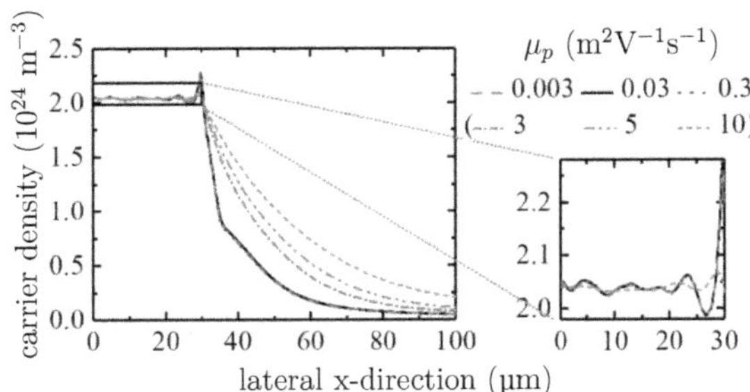

Figure 6.20: Time averaged carrier density distribution at the front facet for different values of the mobility μ_p entering Eq. (3.35). The black solid curve corresponds to the actual hole mobility of the simulated structure (*cf.* Tab. A.6). Unrealistically high values of μ_p are also depicted (marked in the legend by brackets) to visualize the theoretical effects.

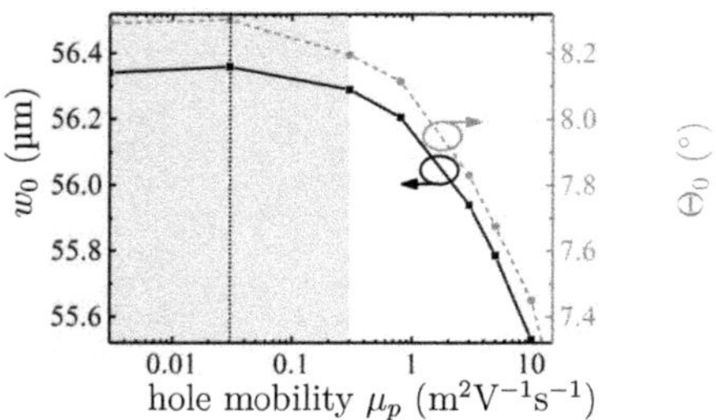

Figure 6.21: Effect of hole mobility on the near-field width and far-field angle. The vertical black-dotted line indicates the actual simulation value of $\mu_p \approx 0.03\ m^2V^{-1}s^{-1}$. A variation of the mobility within one order of magnitude (marked grey) has no influence.

influence. Thus the effect of changes in the hole mobility on spatial hole burning and lateral carrier spreading can be neglected when considering reasonable ranges.

6.3.4 Conclusions

In this section the effect of non-thermal far-field blooming and the dependence of transverse instabilities regarding series resistivity, sheet resistance, mobility, as well as differential index n'_N is investigated. It turns out that a lot of phenomena can be well understood in the mode picture, however this view also has its limitations for example when it comes to anti-index-guided structures.

By neglecting the carrier induced refractive index change $n'_N = 0$ the emitted field is still multi-peaked and shows a high dynamic, suggesting that the induced Kerr-type nonlinearity is not responsible for the multi-peaked near-field structure in BA lasers. By increasing the differential index n'_N more modes contribute to the field and the spatial and temporal scales of the intensity structures become smaller. Distinguishable stable filaments are however not observed even at an extremely high differential index of $n'_N = 2 \cdot 10^{30}\ m^3$ using the traveling-wave model.

To improve the beam quality, higher order lateral modes can be substantially suppressed by implanting the contact layer next to the injection stripe or by lowering the p-layer resistivity. The mobility has only a small impact on the beam quality when considering reasonable ranges.

6.4 Thermal waveguiding effects

In this section the influence of thermal waveguiding on the lateral field distribution is discussed. The formation of a thermally induced waveguide, i.e. a substantial increase of the refractive index in the hot center below the contact stripe is commonly referred to as "thermal lensing". This effect is well studied both experimentally [16] and theoretically [18, 19, 20] for CW operation assuming a stationary temperature distribution.

It is usually neglected under pulsed operation because thermal build-up times up to milliseconds are much longer than the pulse lengths. However, the heat is generated near to the active layer in the same region where the guided wave is localized. This region is small and its thermal build-up time is much shorter than that of the whole device.

The heating model introduced in Chapter 4 takes into account heating processes that take part on small time scales up to some nanoseconds and in Section 6.4.1 exemplary laser operation with 10 ns long pulses is analyzed regarding the influence of small scale heating effects on the field distribution.

The application of the time-dependent approach to CW operation is presented in Section 6.4.2. The influence of spatio-temporal fluctuations of the heat sources is discussed. In a previous publication [17] a time-averaged longitudinally varying temperature profile was obtained with stationary temperatures. Here this longitudinal varying temperature profile is gained self-consistently. The consequent effect of front facet near-field narrowing is discussed and a method for its counteraction presented.

The studied laser has a cavity length of $L = 4$ mm and a width of the contact stripe of $W = 100$ µm (structure II Tab. A.3). To provide lateral current confinement the p-doped layers are considered to be implanted down to the vicinity of the active region, rendering these areas highly resistive. Besides the temperature influence on the real refractive index $\Delta n_T(T)$ explicit thermal changes are only considered in the Fermi function under CW operation. Thermal changes of parameters are disregarded in order to separate those from thermal waveguiding effects.

6.4.1 Short-pulse operation

Lasers emitting high pulse powers with short pulse duration have important applications and are for example utilized in light detection and ranging systems [79] (*cf.* Chapter 5). Thus an investigation of the underlying guiding mechanisms is of interest.

First exemplary laser operation with a 10 ns long pulse is studied by applying the no-heat-flow approximation (*cf.* Section 4.4) at an injection current of 150 A. The bias voltage U is instantaneously turned on and kept constant until it is switched off. After the turn on period, the thermal index generation rate under the stripe (4.44), is non-negative and strongly fluctuating around a mean value of about 50 ms^{-1} as shown in Fig. 6.22. Thus, the refractive index grows like a staircase. The individual steps, however, are typically only picoseconds short and raise the index by less than 10^{-7}, so that their presence does not have an impact on the wave propagation.

The timetrace of the emitted field intensity at the output facet is displayed in Fig. 6.23(a). A qualitative change of the field is visible in this pseudo-color mapping as the field width fluctuates less and seems to become more stable during the transient. This qualitative observation can be confirmed by plotting the evolution of near-field width

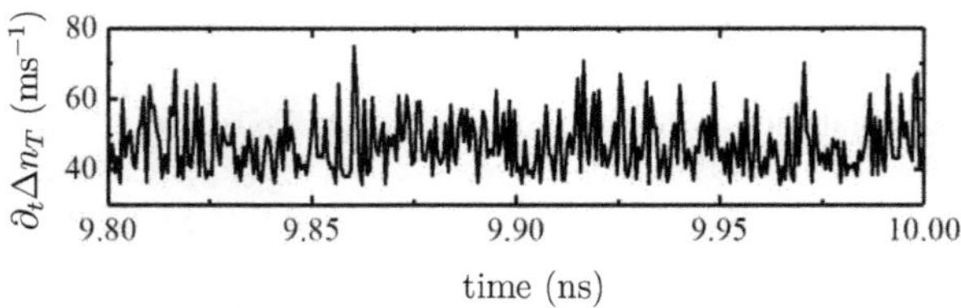

Figure 6.22: Thermal index generation rate of Eq. (4.44) during the last 200 ps of a 10 ns long 150 A current pulse transient.

and far-field angle at the front facet, which were averaged over 1 ns, as their widths decrease gradually during the pulse.

The reason for this is the initialization of a thermally induced waveguide as is visible in Fig. 6.24. The effective index rises below the stripe (red solid in Fig. 6.24) due to the rising thermal contribution, whereas it remains approximately unchanged outside (black dashed). In this process, the index step from outside to inside the stripe $\delta\Delta n_{\text{eff}} = \Delta n_{\text{eff}}(x = 0\ \mu\text{m}) - \Delta n_{\text{eff}}(x = 60\ \mu\text{m})$ rises from negative values immediately after turn on to positive values at the pulse end. At the time of transition, Fig. 6.25(a) shows the lateral profiles of Δn_N and Δn_T together for comparison, averaged over 1 ns. Here the effective index step $\delta\Delta n_{\text{eff}}$ from outside to inside the stripe nearly vanishes, whereas at the pulse end it is clearly positive forming a lateral waveguide as displayed in Fig. 6.25(b). Due to the longitudinal inhomogeneity this transition occurs later at the rear facet, Fig. 6.24(b). In Fig. 6.26(a) and (b), the near and the far-field intensity averaged over 1 ns, are shown. The disappearance of side wings in the near field and a change of the far-field intensity are visible.

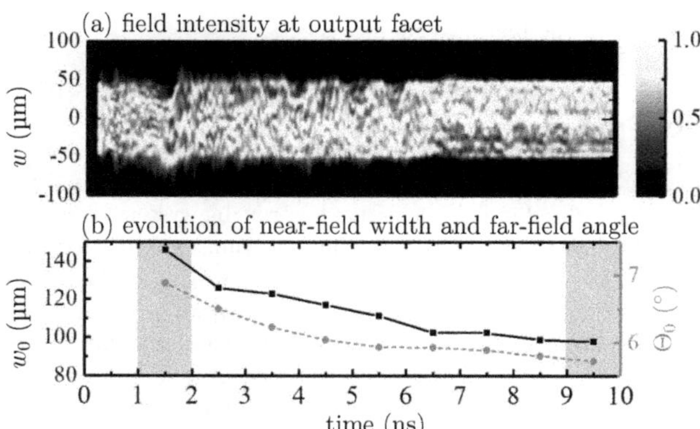

Figure 6.23: Thermal impact on waveguiding for a 10 ns long 150 A current pulse transient. (a) Evolution of field intensity at the output facet. (A moving mean over 25 ps was applied.) (b) Evolution of the near-field width w_0 and far-field angle Θ_0 at front facet, averaged over 1 ns (95% power content). Shadings: Time intervals used in the representations of Figs. 6.25 and 6.26.

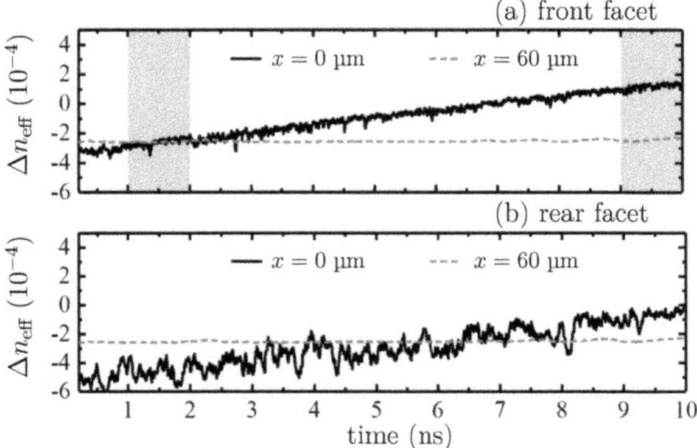

Figure 6.24: Thermal impact on waveguiding for a 10 ns long 150 A current pulse transient. Effective index $\Delta n_{\text{eff}} = \Delta n_N + \Delta n_T$ below ($x = 0$ µm) and beside the stripe ($x = 60$ µm) at (a) front and (b) front facet, respectively. Shadings: Time intervals used in the representations of Figs. 6.25 and 6.26.

In laser structures with nonpositive index steps, lateral optical confinement results from the optical gain and accordingly this regime is usually called gain guided or anti-index-guided. Operation regimes with positive index steps do not require the gain for optical confinement and are referred to as index guided. Thus, Fig. 6.23(a) indicates a transition from gain guiding to index guiding during the pulse. Such a transition has been reported already in [96] explaining long delay times for lasing of narrow-stripe lasers. Fig. 6.23 reveals that this transition is accompanied by a shrinking of both near-field width and far-field angle, which improves the beam quality.

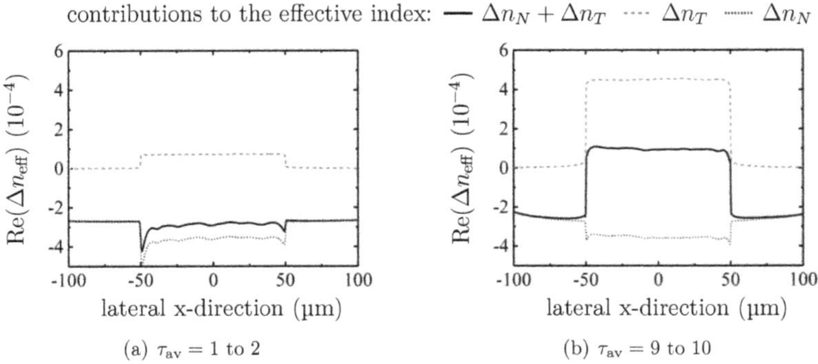

Figure 6.25: Thermal impact on waveguiding for a 10 ns long 150 A current pulse averaged over the pulse within the time interval 1 to 2 ns and 9 to 10 ns (gray shaded in Fig. 6.23(a) and (c)). (a) & (b) Lateral profiles of the effective index. (c) Near and (d) far-field intensities at the front facet.

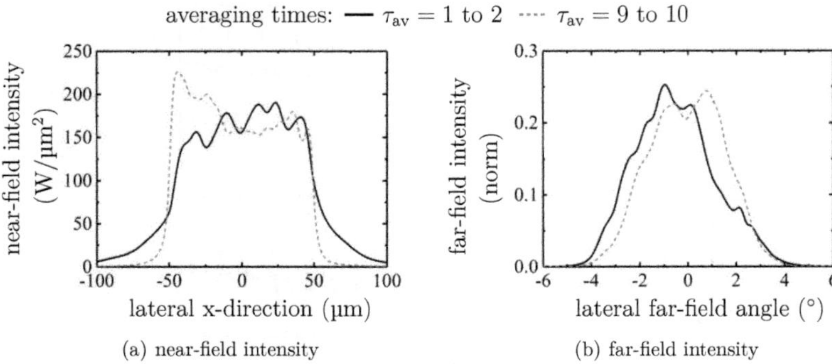

(a) near-field intensity (b) far-field intensity

Figure 6.26: (a) Near and (b) far-field intensities at the front facet for a 10 ns long 150 A current pulse averaged over the pulse within the time interval 1 to 2 ns and 9 to 10 ns (gray shaded areas in Figs. 6.23(b) and 6.24(a))

The described formation of an initial thermally induced waveguide is also detectable in the dependence of time-averaged quantities on the pulse amplitude. Fig. 6.27(b) shows a shrinking of the near-field width and increased far-field angle with increasing injection current (solid curve), compared to the case without thermal waveguiding (dashed). The mean pulse power, given by the total pulse energy divided by the pulse length, as function of pulse current, in contrast, shows no dependence on waveguiding for currents up to 150 A as shown in Fig. 6.27(a). Note, that the neglected changes of parameters with temperature might have an additional influence.

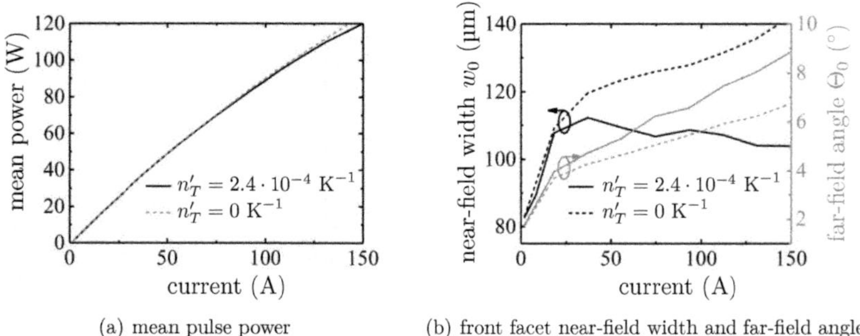

(a) mean pulse power (b) front facet near-field width and far-field angle

Figure 6.27: (a) Mean pulse power and (b) front facet near-field width (black left axis) and far-field angle (red right axis) containing 95 % of the power as function of mean pulse current for 10 ns long pulses with ($n'_T = 2.4 \cdot 10^{-4}$ K^{-1} - solid) and without ($n'_T = 0$ K^{-1} - dashed) thermal waveguiding.

6.4.2 Continuous-wave operation

Whereas under pulsed operation the thermally induced waveguide is growing with increased time of excitation, under CW operation it approaches a steady state as discussed in Section 4.5. Additionally to the much larger magnitude, under CW operation the effective index step below and beside the injection stripe, $\delta\Delta n_{\text{eff}} = \Delta n_{\text{eff}}(x = 0\ \mu\text{m}) - \Delta n_{\text{eff}}(x = 60\ \mu\text{m})$, is a function of the longitudinal coordinate accompanied by a corresponding variation of the width of the lateral field intensity [17], too. This is exemplary displayed in Fig. 6.28 for a total injection current of $I \approx 19.5$ A and power at the output facet of $P_{\text{out}} \approx 19$ W for the same structure discussed in the previous section.

The high asymmetry in the facet reflectivity values gives rise to a strong z-dependence of the field intensity and hence carrier density and Fermi potential due to longitudinal spatial hole burning [97]. This results in a z-dependence of all heat sources as displayed in Fig. 4.4. At this point of operation Joule and absorption heating have the highest fraction of the total heat production ($Q_{\text{Joule}} = 4.0$ W and $Q_{\text{abs}} = 3.9$ W compared to $Q_{\text{rec}} = 2.0$ W and $Q_{\text{defect}} = 1$ W) and as a result the temperature is higher at the front facet compared to the rear facet, Fig. 6.29(a).

The fractured structure of the forward field intensity with small intensity peaks of longitudinal and lateral extent in the range of some micrometers also occur under CW operation, as is exemplary shown in Fig. 6.31(a) for the forward propagating field u^+. Spatial and temporal fluctuations of all heat source terms are the result. However, the fluctuating contribution to thermal waveguiding $\Delta n_{\text{T,fluct}}$ derived from Eq. (4.44) with $h_{\text{fluct}} = h - \bar{h}$ has a negligible influence on the results and can be disregarded under CW operation.

The temperature induced longitudinally varying index profile resulting in a narrowed optical field at the front facet is unfavorable for laser operation due to two reasons: Firstly, a narrowed optical field at the front facet leads to a reduction of the efficiency, because the optical gain is high at the stripe edges due to lateral spatial hole burning which is visible in the lateral profile of the effective gain in Fig. 6.32(a) on the right axis (red solid). Thus the gain is not directly transferred to the radiation field, but rather dissipated via recombination heating. And secondly, because the enhanced power density as visible in Fig. 6.30(a) lowers the threshold for facet damage.

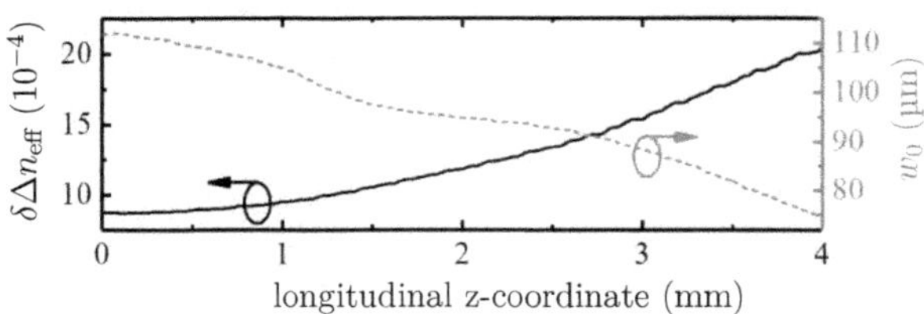

Figure 6.28: CW operation for $I \approx 19.5$ A and $P_{\text{out}} \approx 19$ W. Time-averaged longitudinal evolution of the effective index step below and beside the injection stripe, $\delta\Delta n_{\text{eff}} = \Delta n_{\text{eff}}(x = 0\ \mu\text{m}) - \Delta n_{\text{eff}}(x = 60\ \mu\text{m})$ (left - black axis) and the width (95% power content) of the lateral field intensity (right - red axis).

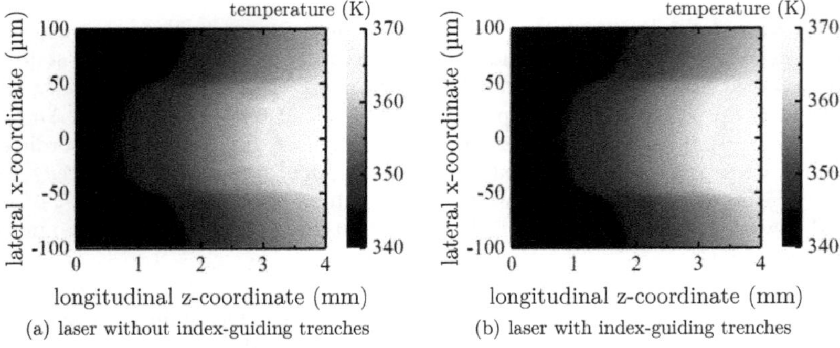

Figure 6.29: Time averaged temperature distribution for the laser without (a) and with (b) index-guiding trenches. CW operation at $I \approx 19.5$ A and $P_{\text{out}} \approx 19$ W.

To provide a temperature and carrier density independent lateral optical confinement, deep index-guiding trenches filled with an insulator can be etched directly next to the injection stripe. Then the term $\Delta n_0(x)$ in (2.28) is negative at the trench position, see the black dashed line in the lateral profiles of the effective index in Figs. 6.32(b) and 6.33(b), where $\Delta n_0 = -1.15 \cdot 10^{-3}$ for $|x| \in [55 : 50]$ µm.

A comparison of the laser operation with and without index-guiding trenches reveals, that although the temperature distribution remains largely unchanged, Fig. 6.29, the near-field narrowing can be successfully counteracted by etching index-guiding trenches, Fig. 6.30.

The lateral profiles $\text{Re}(\langle\Delta n_{\text{eff}}(x)\rangle_t)$ and $2k_0\text{Im}(\langle\Delta n_{\text{eff}}(x)\rangle_t) = \langle g_{\text{eff}}(x) - \alpha_{\text{eff}}(x)\rangle_t$ are displayed in Figs. 6.32 and 6.33 for the front and rear facet, comparing the lasers without and with index guiding trenches. The thermally induced refractive index profile of the laser without index guiding trenches is narrowed at the front facet, because of a narrowed absorption and Joule heating (Fig. 4.4(a) and (b)). For the total refractive

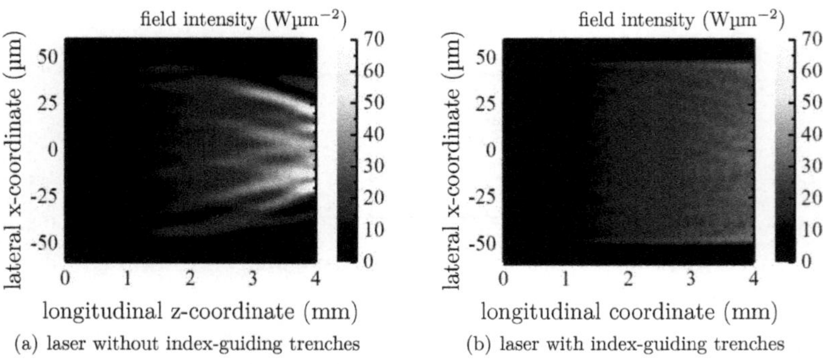

Figure 6.30: Time averaged field intensity distribution for the laser without (a) and with (b) index-guiding trenches. CW operation for $I \approx 19.5$ A and $P_{\text{out}} \approx 19$ W.

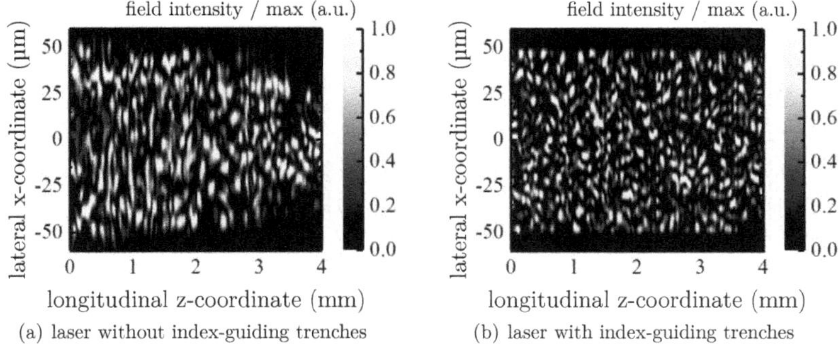

(a) laser without index-guiding trenches (b) laser with index-guiding trenches

Figure 6.31: Forward field intensity distribution at last time instance divided by its maximum at every longitudinal step for the laser without (a) and with (b) index-guiding trenches. CW operation for $I \approx 19.5$ A and $P_{\text{out}} \approx 19$ W.

index change this is even enforced by the carrier induced index change. The refractive index profile of the laser with index-guiding trenches, however, is much wider, because the distribution of absorption and Joule heat source density is more homogeneous (not shown here). Accordingly also the gain function is laterally stabilized and remains box-shaped throughout the device with only small side peaks at the front facet.

The drawback of a laser design with etched index-guiding trenches is a decreased lateral beam quality, i.e. increased beam parameter product $\text{BPP}_{\text{lat}} = w_0\Theta_0/4$. In Fig. 6.34(a) it is visible that the lateral width of the emitted near field w_0 is broadened to the width of the injection stripe. However, the lateral far-field angle Θ_0 remains nearly unchanged, see Fig. 6.34(b), and thus the beam quality is decreased (BPP_{lat} increased). To suggest a reasonable design, it is necessary to counterbalance beam quality against efficiency and reliability as will be discussed in more detail in Section 7.1.

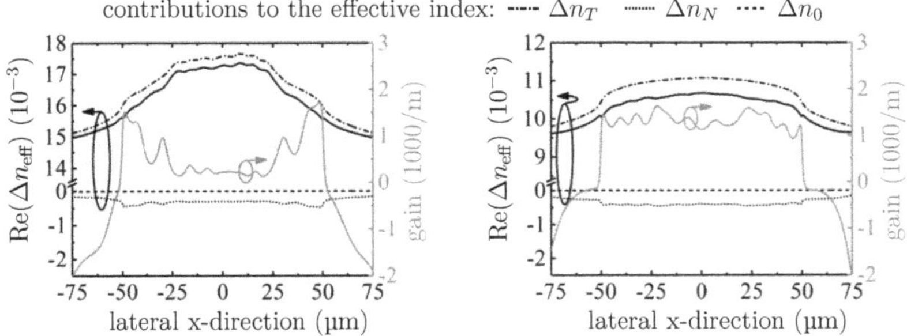

(a) laser without index-guiding trenches (front facet) (b) laser with index-guiding trenches (front facet)

Figure 6.32: Lateral profiles of the time averaged effective index and net gain at the front facet for a laser without (a) and with (b) index-guiding trenches.

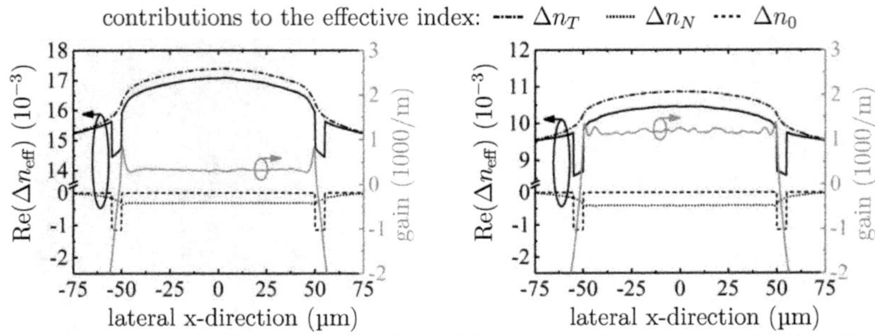

(a) laser without index-guiding trenches (rear facet)

(b) laser with index-guiding trenches (rear facet)

Figure 6.33: Lateral profiles of the time averaged effective index and net gain at the rear facet for a laser without (a) and with (b) index-guiding trenches.

To better understand the reason for the beam quality degradation and difference of the underlying guiding mechanisms involved, the fluctuating complex-valued field profile at the front, $z = L$, and rear, $z = 0$, facets and a given instantaneous time t is expanded in terms of the modes $\phi_m(x)$ of the Helmholtz equation (6.20) for a hypothetical steady state as described in Section 6.2.

The lateral near and far-field intensities of the first six waveguide modes $\phi_m(x)$ having the highest real valued index $\text{Re}(n_m^2)$ obtained from (6.20) are visible in Fig. 6.34(c) for the front facet. In contrast to the modes of the laser with index guiding

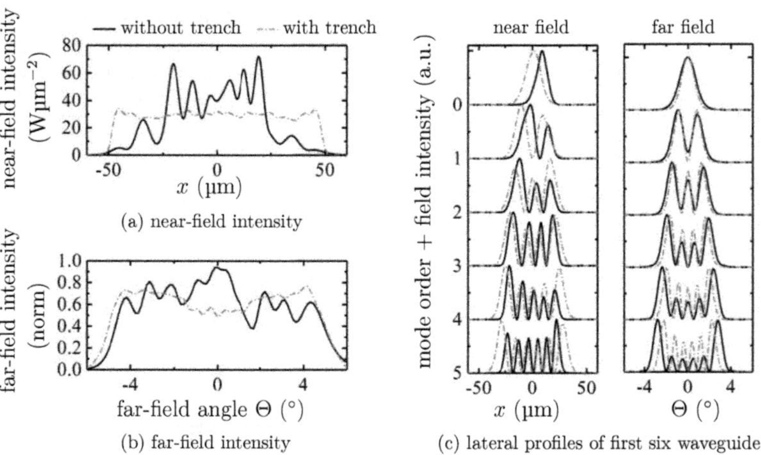

(a) near-field intensity

(b) far-field intensity

(c) lateral profiles of first six waveguide modes $\phi_m(x)$ obtained from (6.20)

Figure 6.34: Comparison of the lasers without (black solid) and with (red-dashed) index-guiding trenches. (a) Time-averaged near and (b) far-field profiles and (c) lateral profiles of the first six waveguide modes $\phi_m(x)$ obtained from (6.20) for the front facet. CW operation for $I \approx 19.5$ A and $P_{\text{out}} \approx 19$ W.

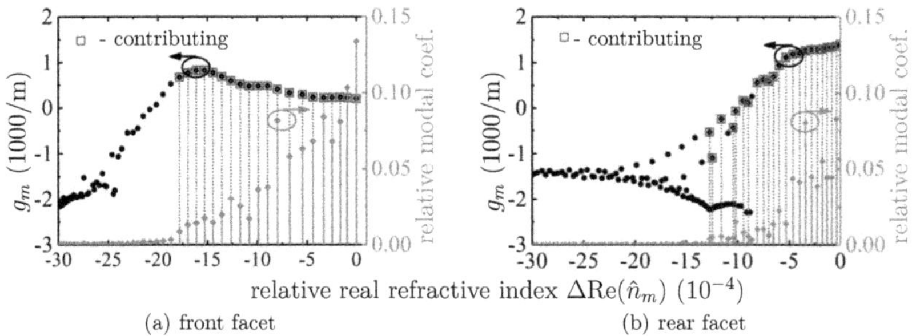

Figure 6.35: Black dots - left axis: Imaginary and real parts of the eigenvalues of (6.20) for the laser without index guiding trenches (a) at front and (b) rear facet. In blue the eigenvalues are indicated that are needed to reproduce the field with a time averaged error (6.23) of $\langle\sigma^+(t,z_0)\rangle_t < 1\%$. Red diamonds - right axis: Corresponding relative modal coefficients $|a_m|^2/\sum_m |a_m|^2$.

trenches in the case without index guiding trenches the modal near-field intensities are often asymmetric, because the field is more vulnerable to fluctuations in the effective gain and refractive index. Interestingly, although the waveguide mode far fields of the laser without index guiding trenches are wider, the actual far fields of the lasers with and without index guiding trenches have the same width.

This can be understood by examining the relative modal coefficients $|a_m|^2/\sum_m |a_m|^2$ obtained from the decomposition of the output field according to Eq. (6.22). For the lasers with and without index-guiding trenches the imaginary and real parts of the eigenvalues of (6.20) are displayed in Fig. 6.35 and Fig. 6.36 as black dots, respectively. In blue those eigenvalues are indicated that are needed to reproduce the field with a

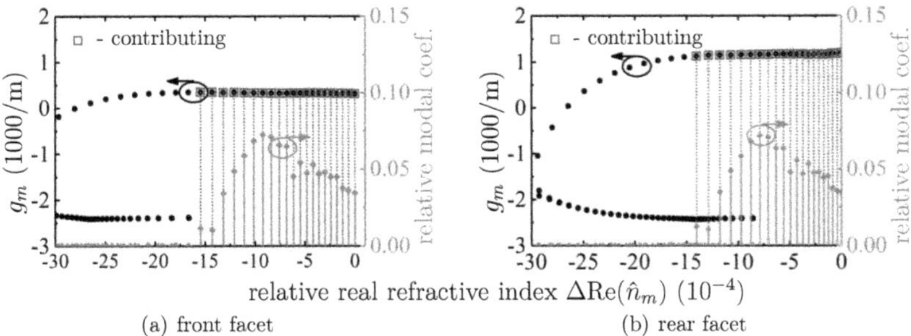

Figure 6.36: Black dots - left axis: Imaginary and real parts of the eigenvalues of (6.20) for the laser with index guiding trenches (a) at front and (b) rear facet. In blue the eigenvalues are indicated that are needed to reproduce the field with a time averaged error (6.23) of $\langle\sigma^+(t,z_0)\rangle_t < 1\%$. Red diamonds - right axis: Corresponding relative modal coefficients $|a_m|^2/\sum_m |a_m|^2$.

time averaged error (6.23) of $\langle\sigma^+(t,z_0)\rangle_t < 1\%$. Furthermore the magnitudes of the relative modal coefficients $|a_m|^2/\sum_m |a_m|^2$ are plotted as red diamonds.

In the case of the laser without index-guiding trenches the modal gain peaks at higher order modes, Fig, 6.35(a) due to carrier accumulation at the edges, Fig. 6.32. Still their contribution to the field remains relatively low compared to low order modes. In total 20 modes are necessary to reproduce the field at the front facet, whereas at the rear facet a total number of 26 modes are needed. In the case of the laser with index-guiding trenches, however, 21 modes are needed to reproduce the field, at both, rear and front facets. The contributing modes all nearly have the same modal gain, whereas the modes with the highest contribution to the field are of a higher order. The excitation of higher order modes explains why the far-field angle of the lasers without and with index-guiding trenches are the same although the modal far fields of the laser without index-guiding trench are smaller.

6.4.3 Conclusions

In this section the influence of slow and fast contributions to the time-dependent temperature on the lateral near- and far-field distributions are investigated.

The simulation of an exemplary laser operation with 10 ns long pulses reveals a fast-growing thermally induced waveguide initializing the known stationary thermal lense. Although the temperature increase is small, a waveguide that is introduced within a few-nanoseconds-long pulse can result in a transition from a gain-guided to an index-guided structure, leading to a near-field narrowing.

The application of the time-dependent approach to CW operation yields strong spatio-temporal sub-nanosecond fluctuations of the heat sources. However, they have a negligible influence on the wave propagation, whereas the effects of the time-averaged londitudinally varying temperature profile confirm those obtained with stationary temperatures in a previous publication [17]. It results in a near-field narrowing which is unfavorable, firstly because of carrier accumulation at the stripe edges and secondly because the enhanced power density lowers the threshold for facet damage.

It is shown how these effects can be tailored by appropriate lateral index-guiding trenches, which, however, results in a beam quality degradation as higher order modes have a higher contribution in the lasing process.

Chapter 7

Improvement of the lateral brightness

For application, both, a high output power P_{out} and good beam quality, i.e. small beam parameter product BPP_{lat}, are needed, which are both included in the target figure brightness, $B_{\text{lat}} = P_{\text{out}}/\text{BPP}_{\text{lat}}$. The poor lateral beam quality of BA lasers has been a long standing problem and thus different approaches for beam quality improvement have been proposed in the past, amongst them are methods for tailoring of the current [22, 98, 99] or thermal path [100, 101, 102] or structuring of the contact region [103, 104, 105, 106, 107].

In Section 6.3 the effects that lead to far-field broadening under pulsed operation have already been discussed in detail. From here conclusions for the improvement of the lateral beam quality can be directly drawn: Higher order lateral modes can be substantially suppressed by implanting the contact layer next to the injection stripe or by lowering the p-layer resistivity.

Under continuous-wave operation (Section 6.4.2) etching index-guiding trenches leads to a degradation of the beam quality whereas it counteracts the temperature induced near-field narrowing and thus improves the efficiency and possibly lowers the threshold for catastrophic optical damage (COD) of the laser. This observation suggests that a maximum brightness can be found by varying trench parameters such as the effective index step, trench width and distance of the trench to the injection stripe to find the optimum trade-off between efficiency and beam quality. The first part of this chapter focuses on this question. A further impact of implantation will be discussed as well and simulation results are compared with measurements.

In order to achieve a high far-field peak at 0° lateral far-field angle the contact region can be structured with injection stripes in the range of some micrometer width with some micrometer spacing in between. If the emitted field below each injection stripe of such arrays lasers [104] is co-phasal a diffraction pattern similar to the one that is achieved for an optical grating is visible. However, achieving in-phase operation is difficult and thus mode-selection mechanisms have to be employed.

In the second part of this chapter coherently coupled laser arrays and mode-selection mechanisms to achieve in-phase operation are discussed. In particular a novel laser type is presented that has an additional longitudinal structuring of the injection stripe and index-guiding trenches for phase tailoring to achieve in-phase operation at extremely high output powers under pulsed operation.

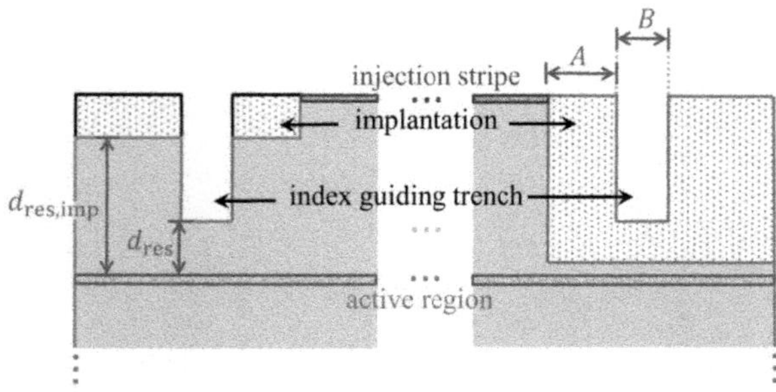

Figure 7.1: Sketch of the shallow implanted (left) and deep implanted (right) laser device. The index-guiding trenches have a distance A to the injection stripe and width of B. d_{res} denotes the distance of the index-guiding trench to the active region and $d_{res,imp}$ the distance of the deepest implanted part to the active region.

7.1 Index guiding trenches and implantation

The laser examined in this section has a an injection stripe length of $L = 4$ mm and width of $W = 100$ µm (structure II Tab. A.3). The simulation was performed using the thermal model (see Section 4.5). The simulation time is $\tau_{sim} = 8$ ns and over the last $\tau_{av} = 5$ ns the time averaged data is obtained. For the calculation of the temperature profile at least 3 iterations are performed. Only the thermally induced waveguide is considered in the calculations. The neglected changes of parameters with temperature might have an additional effect.

To find the optimum trade-off between beam quality and durability and efficiency the distance of the trench to the injection stripe A (see Fig. 7.1) can be varied. This influences the near-field width and far-field angle as is visible in Fig. 7.2 for a built-in index step of $\Delta n_0 = -1.15 \cdot 10^{-3}$ at an output power of $P \approx 20$ W. Additionally the

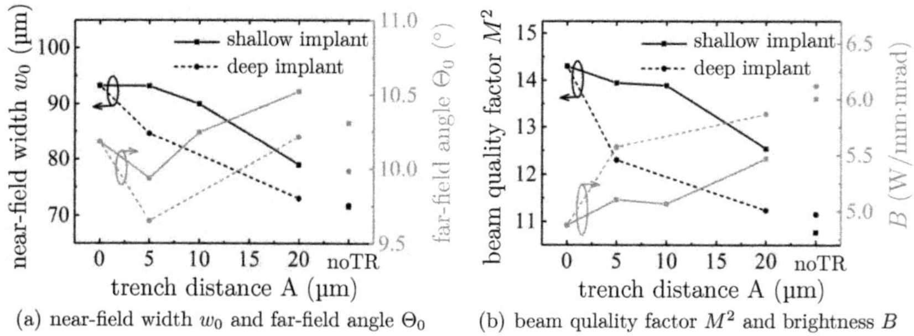

(a) near-field width w_0 and far-field angle Θ_0 (b) beam qulality factor M^2 and brightness B

Figure 7.2: Near-field width (left black axis) and far-field angle (right red axis) as function of the trench distance for a built-in index step of $\Delta n_0 = -1.15 \cdot 10^{-3}$ and constant voltage of $U = 1.6$ A corresponding to $I \approx 20$ A and $P \approx 20$ W. Shallow ($d_{res,imp} = 1610$ nm) and deep ($d_{res,imp} = 50$ nm) implanted structures are shown for comparison.

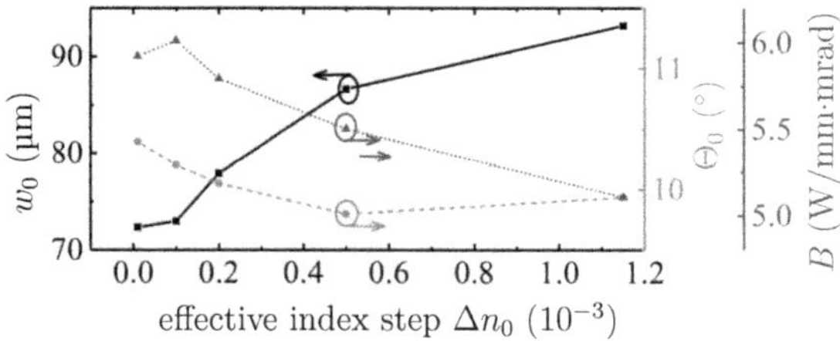

Figure 7.3: Lateral near-field width and far-field angle and lateral brightness for a constant voltage of $U = 1.6$ A corresponding to $I \approx 20$ A and $P \approx 20$ W.

structure can be implanted to suppress the excitation of higher order lateral modes by reducing current spreading and lateral carrier accumulation, see Fig. 7.1. The shallow implanted laser refers to a contact implantation with a residual layer thickness of $d_{\text{res,imp}} = 1610$ nm (marked in Fig. 7.1). The deep implanted laser has an implanted residual layer thickness of $d_{\text{res,imp}} = 50$ nm.

Reducing the distance of trench to injection stripe has a similar effect as a deep implantation as the beam quality and brightness increase (beam quality factor M^2 decreases). Enlarging the trench width from 5 to 20 µm has only a small impact (not shown here).

Comparing the laser without trench distance to the injection stripe ($A = 0$ µm) to the laser without trench, the near-field width is increased by 30% from ≈ 70 to ≈ 90 µm and the brightness is degraded by 17% from 6 to 5. At 5 µm trench distance and a shallow implantation the near-field width is still widened above 90 µm, but a slight decrease in the far field is visible, leading to a slight brightness improvement.

For this case ($A = 5$ µm, $B = 5$ µm, $d_{\text{res,imp}} = 1610$ nm) Fig. 7.3 shows the influence of an effective index variation on the near-field width w_0, far-field angle Θ_0 and brightness B. Here only the optical effect of the trench has been taken into account and electrically only a contact implantation ($d_{\text{res,imp}} = 1610$ nm) is considered. A similar influence as observed in Fig. 7.2 can be seen: Decreasing the effective index step from $\Delta n_0 = -1.15 \cdot 10^{-3}$ to 0 leads to a decrease in the near-field width and improvement in brightness. The maximum brightness is obtained at an effective index step near $\Delta n_0 = -10^{-4}$, whereas still a good brightness and broad near field is gained at $\Delta n_0 = -5 \cdot 10^{-4}$.

7.1.1 Comparison with measurements

The measured and simulated PI characteristics of a laser without and with index-guiding trench ($A = 5$ µm, $B = 5$ µm, $\Delta n_0 = -1.15 \cdot 10^{-3}$) and contact implantation ($d_{\text{res,imp}} = 1610$ nm) with length $L = 6$ mm and width $W = 90$ µm are shown in Fig. 7.4 as solid and dashed lines, respectively. For the measurement the laser chips were soldered p-side down on CuW submounts. The measurements were performed under CW operation with a heat sink temperature of $T = 300$ K.* The mean values of two laser diodes are depicted.

The simulation is performed in the same way as described above, however in contrast to the previous investigation the parameter dependence on temperature is included

*The author thanks M. Beier and K. Häusler for the CW measurements.

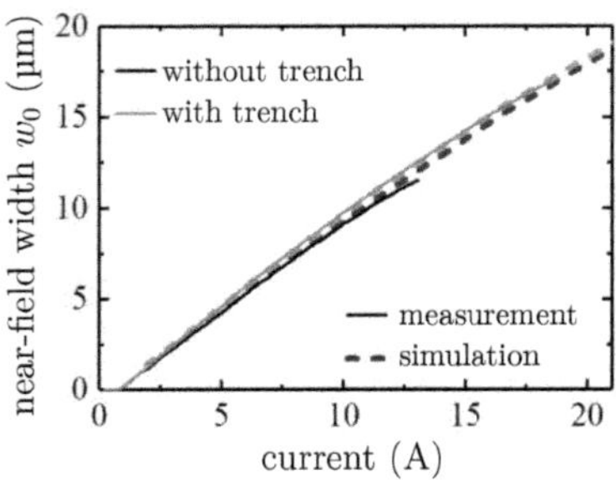

Figure 7.4: Comparison of the measured and simulated PI characteristics of BA lasers without and with index-guiding trenches (A = 5 µm, B = 5 µm, $\Delta n_0 = -1.15 \cdot 10^{-3}$). Investigated lasers had a shallow implantation ($d_{\mathrm{res,imp}} = 1610$ nm).

in the calculations (see Appendix A). For the simulation the parameters for internal background absorption α_0, the Shockley Read Hall recombination coefficient A and the thermal transmission resistance r_{th} have been adjusted to fit the PI slope at small currents, the PI threshold and the simulated active region temperature below the injection stripe and near the front facet, respectively, (see Tab. A.5 compared to Tab. A.3). The adjusted temperature was experimentally determined by a fit of the wavelength shift with current of $\Delta\lambda/\Delta I \approx 0.7$ nm/A with $\Delta\lambda/\Delta T = 0.33$ nm/K known from independent measurements and a slope the power-current characteristics of $\Delta P/\Delta I = 1.1$ W/A which results in a temperature increase with output power of $\Delta T/\Delta P = 2$ K/W.

The measured PI curve of the laser without index-guiding trenches shows a stronger bending compared to the laser with index-guiding trenches, Fig. 7.4. The simulation also predicts this behavior and the agreement between measurement and simulation is very good. It shall be pointed out here that both lasers without index-guiding trenches showed COD at a much lower injection current of I = 13 A than the lasers with index-guiding trenches at $I = 19$ A.

In Figs. 7.5 and 7.6 the measured and simulated near- and far-field intensities are displayed at an output power of $P = 10$ W for comparison. The general agreement of measured and simulated near-field intensities is good for the laser with index-guiding trenches, Fig. 7.5(b). Both experimental and simulated near fields of the laser without index-guiding trenches, Fig. 7.5(a), are narrower than in the case of the laser with index-guiding trenches, however, the measured near-field width is broader than predicted by the simulation. In both cases the measured far field, Fig. 7.6, is much broader than predicted in the simulation.

Plotting the near-field width containing 95 % of the power as functions of current, Fig. 7.7(a), shows that the narrow near field of the gain guided laser measurements is already visible at small currents and thus is not the result of a thermally induced waveguide. This is not predicted by the simulation were the near field broadens first with current and then is abruptly narrowed at currents larger then 10 A.

The narrowed near field is the reason for the reduced slope of the measured and simulated PI curve, Fig. 7.4, because a larger part of the carriers do not take part in stimulated recombination as also discussed in Ref. [17]. Furthermore the early COD

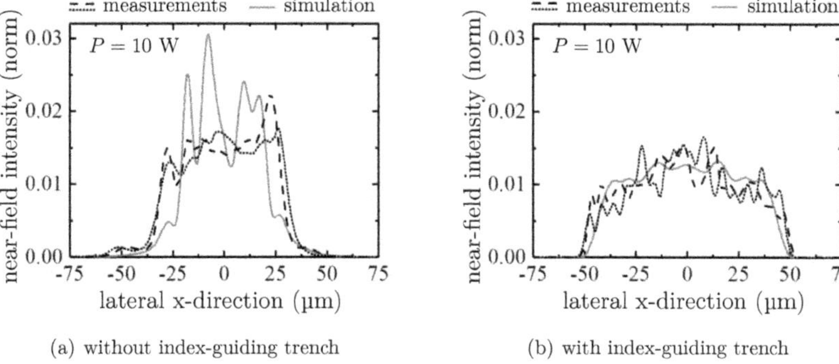

(a) without index-guiding trench

(b) with index-guiding trench

Figure 7.5: Measured and simulated normed near-field intensity $\mathrm{NF}(x)/\int \mathrm{NF}(x)dx$ for the lasers without (a) and with (b) index-guiding trench at an output power of 10 W.

of the gain guided laser can be explained in this way. Estimating the near-field width at COD with $w_0 = 70$ and 85 µm and output power of $P_{\mathrm{out}} = 11.6$ and 17.1 W for the laser without and with index-guiding trenches, respectively, gives a similar field intensity at COD of $P_{\mathrm{out}}(w_0 d_{1/e^2})^{-1}$ of 0.11 W/µm^2 and 0.13 W/µm^2, respectively, with a vertical near-field width of $d_{1/e^2} = 1.5$ µm.

Whereas for small current values, the far-field angle is broader for the laser with index-guiding trenches, Fig. 7.7(b), at currents larger 10 A it becomes smaller compared to the far-field angle of the gain guided laser. The simulated far-field angle is predicted to be smaller compared to the measurements for both laser types. The overall development as function of current, however, is very well predicted.

The same is true for the behavior of beam quality factor and brightness with current, Fig. 7.8. Due to the constant off-set of the simulated compared to the measured far-field angles, beam quality and brightness show a constant off-set as well, Fig. 7.8, but the overall behavior is correctly predicted. The experimental brightness, Fig. 7.8(b) of the laser with index-guiding trenches is always smaller for all injected current values

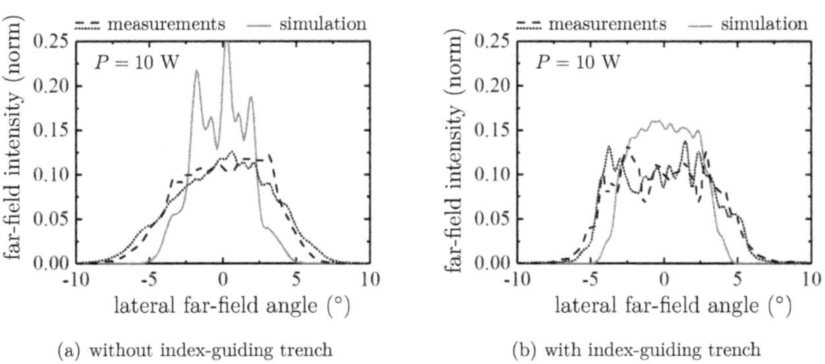

(a) without index-guiding trench

(b) with index-guiding trench

Figure 7.6: Measured and simulated normed far-field intensity $\mathrm{FF}(x)/\int \mathrm{FF}(x)dx$ for the lasers without (a) and with (b) index-guiding trench at an output power of 10 W.

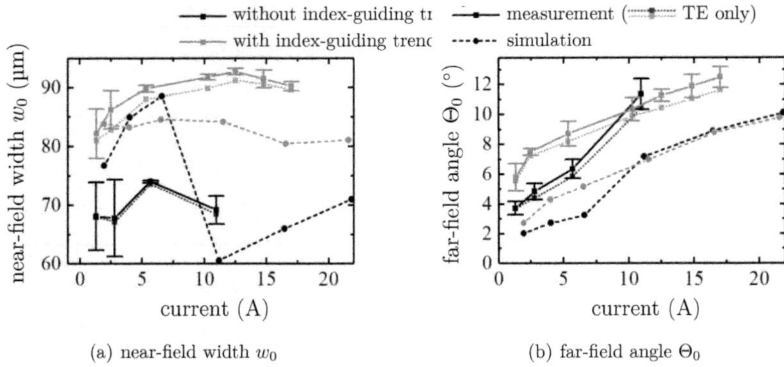

(a) near-field width w_0

(b) far-field angle Θ_0

Figure 7.7: Comparison of the measured and simulated (a) near-field width w_0 and (b) far-field angle Θ_0 with increasing current for a BA laser without and with index-guiding trenches. Results for the measured TE-polarized contributions are shown for comparison. Investigated lasers had a shallow implantation ($d_{\mathrm{res,imp}} = 1610$ nm).

compared to the gain guided laser. However, the slope of the brightness with current remains constant, whereas in the case of the gain guided laser it saturates, both in the simulation and experiment. Hence, it is possible that at high injection currents the brightness of the laser with index-guiding trench exceeds the one of the laser without trench.

The deviation of measured and simulated far-field angles is already visible at small currents. This suggests, that the reason for the deviation is not only a result of an incorrect description of thermal waveguiding, but has a different origin. Some possibilities are discussed in the following.

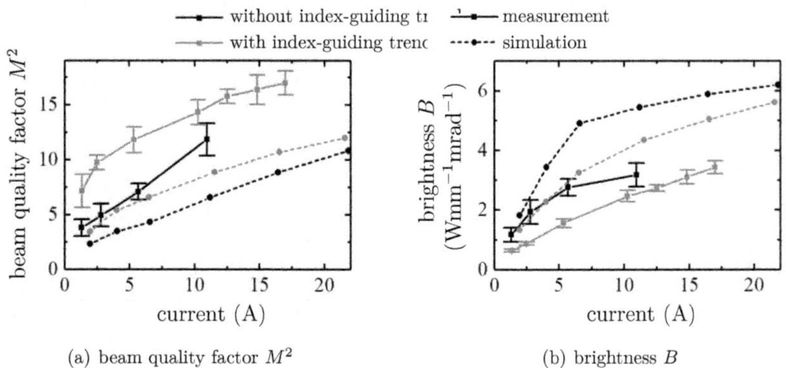

(a) beam quality factor M^2

(b) brightness B

Figure 7.8: Comparison of the measured and simulated (a) beam quality factor M^2 and (b) brightness B with increasing current for a BA laser without and with index-guiding trenches. Investigated lasers had a shallow implantation ($d_{\mathrm{res,imp}} = 1610$ nm).

Field intensities have been calculated in the paraxial approximation, i.e. the second derivative in longitudinal direction is neglected ($\partial_z^2 u^\pm = 0$). This approximation is good because, $\partial_z^2 u^\pm \ll 2\bar{n}k_0\partial_z u^\pm$, holds for the structures investigated in this section and throughout this thesis, so that the discrepancy of measurement and simulation can't be explained by this.

Furthermore, in the Ansatz (2.1) the optical field is represented by its transverse-electric (TE) x-component. Shear strain, which can be induced by build in stress, is not contained in the model and can cause a rotation of the polarization axis of the optical field, so that an electric field vector that initially oscillates in the active region plane (TE-polarized) will be rotated according to the permittivity tensor of the material [108]. To check the accuracy of this approximation in Fig. 7.7 the near-field width and far-field angles of the measured TE contribution are displayed for comparison. The measurement was performed by inserting a polarization filter into the optical path so that only light passes where the electric field component oscillates in the active region plane. It is visible that those contributions indeed show near and far fields that are by up to 6.5% narrower than the total fields. Still, the deviation is small compared to the difference of the measured and simulated field and is not alone responsible for the discrepancy between simulation and experiment.

In [108] the dependencies of the effective refractive index on carrier density $n_N(N)$ and temperature $n_T(T)$ have been used to fit the experimentally determined near and far fields resulting in a very good agreement of simulation and measurements. Besides the fact that the behavior of $n_N(N)$ and $n_T(T)$ is similar in this work compared to [108], the good agreement is a result of the strongly filamented carrier density due to the assumption of field stationarity. Although this assumption is highly doubtful, a stronger lateral variation of the carrier density than predicted here could be the reason for the discrepancy between simulation and experiment. Stronger lateral spatial hole burning is for example a result of a higher series resistance r_s, however also for a constant injection current density ($r_\mathrm{s} \to \infty$) a lateral carrier density modulation of the magnitude predicted in [108] does not arise using the time-dependent traveling wave model (*cf.* Fig. 3.4(c)).

7.1.2 Conclusions

By changing lateral design parameters (implantation depth, built-in index step, trench width and distance of the trench to the injection stripe) a maximum brightness can be found. However, this brightness is only marginally higher than the brightness obtained without index-guiding trenches, so that regarding beam quality improvement etching index-guiding trenches is counterproductive. However, if the COD threshold of the laser and its efficiency are taken into account, etching index-guiding trenches has a positive effect, which is supported by measurements. Due to brightness saturation of lasers without index-guiding trenches it is thinkable that the brightness of lasers with index-guiding trenches exceeds the brightness of the former at high injection currents.

The main discrepancy between measurements and simulation is that the simulated far-field angle is much smaller than measured. Still, the overall trend of lateral far-field angle and brightness as function of current is correctly predicted. A higher modulation of the lateral carrier density distributions and the resulting interaction with the optical field via the carrier density induced refractive index could lead to a broader lateral far field. However, also for a constant injection current density a lateral carrier density modulation of the necessary magnitude does not arise using the time-dependent traveling wave model.

7.2 Contact structuring

Filtering modes by adequate structuring of the contact region in order to improve beam quality have been proposed in numerous publications [103, 104, 105, 106, 107]. Here, in the first two sections the focus lies on coherently coupled laser arrays and the Talbot-type spatial filter, that utilizes properties of near-field diffraction to obtain a small far-field divergence.

In both cases the broad lateral aperture of the laser device, that would otherwise support a large number of lateral modes, is electrically separated into smaller injection stripes by e.g. implantation. Within the narrow injection stripes preferably only the fundamental mode is excited. If the laser array couples incoherently the resulting far field would be broad. In order to achieve high brightness in coherently coupled arrays, an array laser has to be operated in-phase which means that the fields in each element are co-phasal, contrary to out-of-phase when fields in adjacent elements have a π phase-shift. Then the resulting far field shows a narrow single lobe around 0° far-field angle.

The results found in literature are compared to simulations of the traveling-wave model in order to show the progress leading to the newly developed chessboard laser in the third section. It bases on the findings of [109] which shows that longitudinal-lateral gain-loss modulation with periods corresponding to the Talbot length result in an anisotropic gain. In this work, it is shown how the out-of-phase mode can be successfully suppressed by additional phase tailoring, so that extremely high brightness is predicted theoretically for the central lope under pulsed operation.

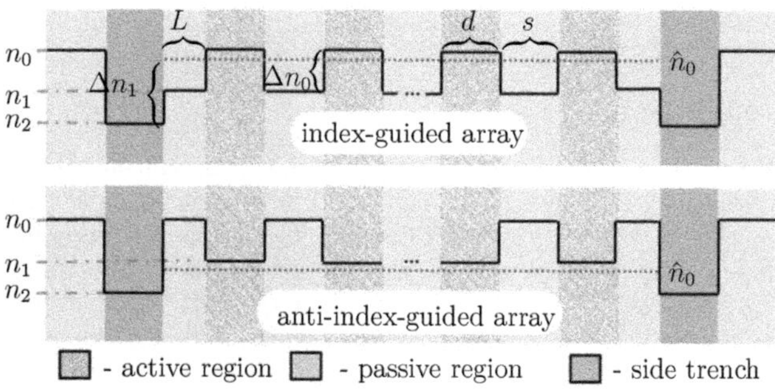

Figure 7.9: Sketch of the real lateral index distribution in cold array lasers. Distinguished are the two cases of the index-guided array, where the mode resides in the high index regions and anti-index-guided arrays, where the mode resides in the low index regions. For mode selection the active regions (dashed areas) are chosen accordingly. At the array edges side trenches (dark grey area) are added. Δn_0 and Δn_1 denote the build in index variations between the array elements and of the index-guiding trenches, respectively, and $\hat{n}_0$ the modal indices of the fundamental mode.

7.2.1 Coherently coupled laser arrays

Array lasers consist of narrow regions of carrier injection (of width d) that are alternated with interelement passive regions (of width s), see Fig. 7.9. In the same way, the real lateral index distribution is alternating with regions of high index n_0 and regions of low index n_1. In array lasers two types of array modes are of importance, the index and anti-index-guided modes. Due to their behavior in the low index regions, index-guided modes are sometimes referred to as evanescent-type modes and have an eigenvalue between n_0 and n_1, whereas anti-index-guided modes are sometimes referred to as leaky-type modes and have an eigenvalue below n_1 [110]. In laser arrays mode selection can be obtained by placing gain in the high index regions to excite index-guided modes or by placing it in the low index regions to excite anti-index-guided modes, see Fig. 7.9.

As index-guided array lasers are more easily fabricated, those were historically studied first. However, simple index-guided arrays without additional mode-selection mechanisms will preferably lase on the out-of-phase mode. To illustrate this in the following a 10 stripe array with stripe width of $d = 4$ µm and interelement distance $s = 2$ µm resulting in a lateral period of $\Lambda_x = 6$ µm with an interelement built-in index step of $\Delta n_0 = -2 \cdot 10^{-2}$ and trenches at the section edges with a built-in index step of $\Delta n_1 = -5 \cdot 10^{-2}$ is studied. For calculation of the gain a constant carrier density of $3 \cdot 10^{24}$ m^{-3} was assumed below the contact stripes.

In Fig. 7.10 the near-field amplitudes and far-field intensities of the 10 highest gain modes obtained from the solution of the Helmholtz equation (6.20) are shown. Due to the finite width of the injection region the individual array amplitudes are modulated by an envelope function corresponding to modes of a potential well. The in-phase mode is the mode of order 0, where order denotes the number of the near-field amplitudes

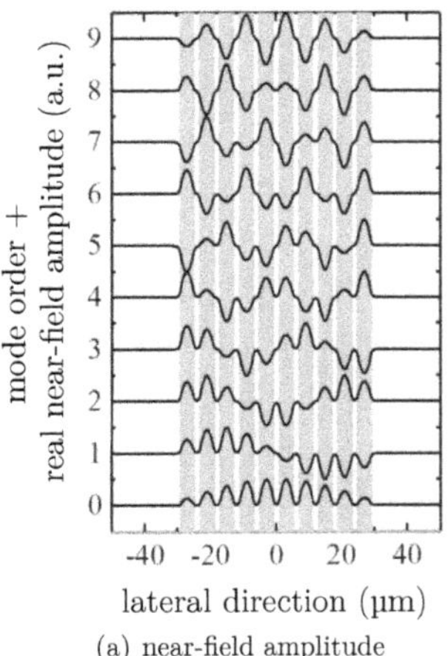

(a) near-field amplitude

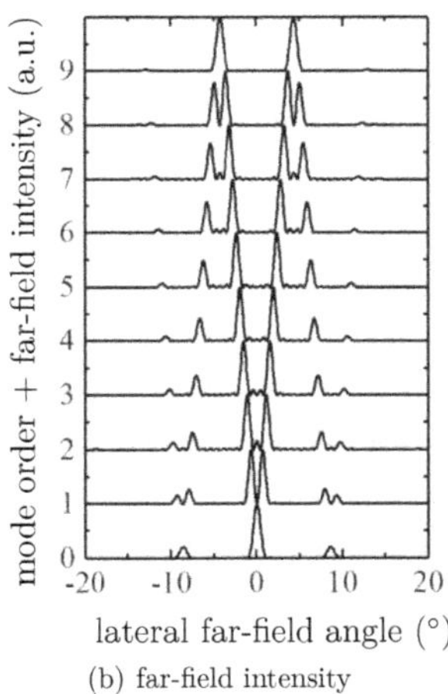

(b) far-field intensity

Figure 7.10: (a) Near-field amplitudes and (b) far-field intensity in a 10 stripe index-guided laser array with stripe widths of $d = 4$ µm and interelement distances of $s = 2$ µm obtained from the solution of the Helmholtz equation (6.20). The grey shaded area in (a) indicates regions of high imaginary and real parts of the effective index.

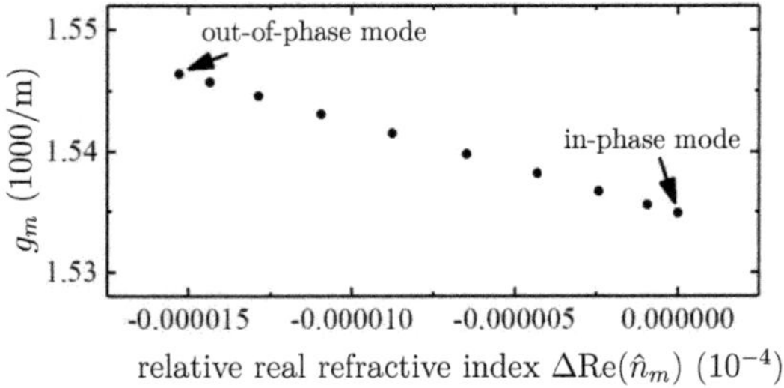

Figure 7.11: Real (abscissa) and imaginary (ordinate) eigenvalues of the modes obtained from the solution of the Helmholtz equation (6.20) an index-guided laser array.

zero transitions. It shows one distinct central peak in the far field with width $\lambda_0/(9\Lambda_x)$ and side peaks at $\pm\lambda_0/\Lambda_x$. The out-of-phase mode is the mode of order 9 that shows distinct side peaks in the far field at $\pm\lambda_0/(2\Lambda_x)$.

As the amplitude of the in-phase mode has no zero transitions in the passive area between the injection stripes, it generally has a higher loss than the other modes. This is also illustrated in Fig. 7.11 where the ordinate is the modal gain, $g_m = 2k_0\mathrm{Im}(\hat{n}_m)$, and the abscissa the real part of the modal index relative to the real part of the index of the fundamental mode, $\Delta\mathrm{Re}(\hat{n}_m) = \mathrm{Re}(\hat{n}_m) - \mathrm{Re}(\hat{n}_1)$. As a result the out-of-phase mode will lase preferably.

As in-phase operation can't be achieved in simple index-guided laser arrays without any additional mode-selection mechanism, the research interests shifted to anti-index-guided array lasers [111, 104, 112]. The simplest anti-index-guided laser is a pulsed operated laser without index-guiding trenches, as the carrier density induced index change leads to anti-index guiding. Anti-index-guided lasers with higher effective index steps between elements are more difficult to fabricate, because the regrowth in a two-step epitaxy is necessary [110].

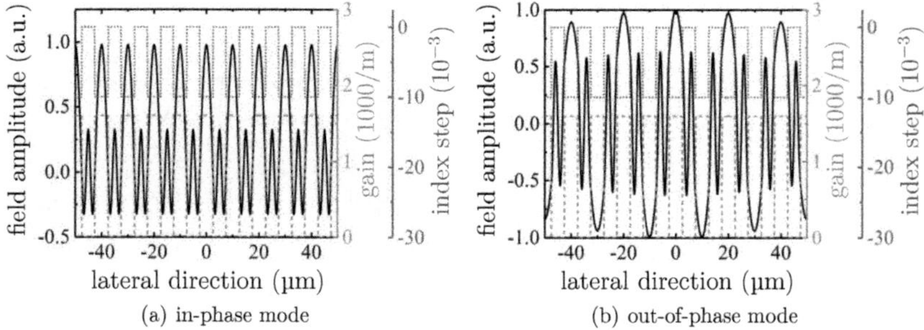

Figure 7.12: Cut-out of the near-field amplitudes of the (a) in-phase and (b) out-of-phase mode in a 27 stripe anti-index-guided laser array with a stripe width of $d = 5$ µm, interelement distance $s = 5$ µm and built-in index step of $\Delta n_0 = -1 \cdot 10^{-2}$.

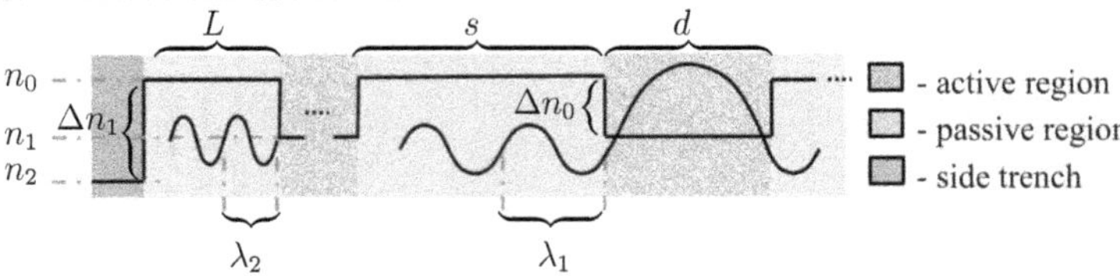

Figure 7.13: Sketch of the real lateral index distribution of an anti-guided laser array and near-field amplitude profile. λ_1 and λ_2 denote the periodicity of the anti-guided mode in the high-index and side trench regions, respectively.

In order for anti-index-guided laser array elements to couple resonantly in-phase or out-of-phase the following resonant condition has to be met for the mode periodicity in the high index regions (see Fig. 7.13) $\lambda_1 = \lambda_0/(\sqrt{n_1^2 - n_0^2 + (\lambda_0/2d)^2})$, with

$$s = m\lambda_1/2 \quad \text{for} \quad \begin{array}{lcl} m & = \text{odd} & \rightarrow \quad \text{in-phase} \\ m & = \text{even} & \rightarrow \quad \text{out-of-phase,} \end{array} \tag{7.1}$$

where m is the number of interelement near-field intensity peaks [104]. Condition (7.1) has been complemented by an additional condition for the distance L of edge index-guiding trench to the injection array (see Fig. 7.13),

$$L = \frac{(p\pi + \varphi)\lambda_2}{2k_0} \tag{7.2}$$

$$\text{with} \quad \lambda_2 = \frac{\lambda_0}{\sqrt{n_0^2 - n_1^2 + (\lambda_0/2d)^2}} \quad \text{and} \quad \varphi = 2\tan^{-1}\left(\frac{\sqrt{n_1^2 - (\lambda_0/2d)^2 - n_2^2}}{\sqrt{n_0^2 - n_1^2 + (\lambda_0/2d)^2}}\right) \tag{7.3}$$

with the integer number p [113].

Assuming isothermal conditions and neglecting the carrier density induced index change and $d = s = 5$ µm results in an effective index step of $\Delta n_0 = -1 \cdot 10^{-2}$ for $m = 3$ and with $p = 3$ and $\Delta n_1 = 2 \cdot 10^{-2}$ a distance of the edge index-guiding trench to injection array of $L = 2.1$ µm is gained. The corresponding near and far field of the in-phase and out-of-phase mode of this structure are shown in Fig. 7.12. Here a carrier density of $N = 3 \cdot 10^{24}$ m^{-3} was assumed, as in the preceding case. In this case the highest gain is obtained for the in-phase mode.

For the simulation of the anti-index-guided laser array with the traveling wave model structure II was used. The simulation was performed without temperature and for a simulation time of 10 ns, during which the voltage was suddenly set to $U = 1.55$ V (resulting in $I = 1$ A), respective $U = 2.0$ V (resulting in $I = 10$ A), and kept constant until end. The time-averaged fields displayed in Fig. 7.14 are obtained by averaging over the last 5 ns of this time period. Using the traveling wave model a diffraction-limited far-field profile was only obtained directly above threshold at $I \approx 1$ A, Fig. 7.14(a). For higher currents lateral spatial hole burning, Fig. 7.14(b), results in a reduced beam quality.

To improve the stability a higher Δn_0 can be chosen as proposed in [110]. Still, full coherence and at the same time stability is difficult to obtain. In e.g. [114] 11.5 W laser operation with a far field of a central lobe full width at half maximum (FWHM) of 1° is presented, however, with a comparatively low single lobe power content. Even

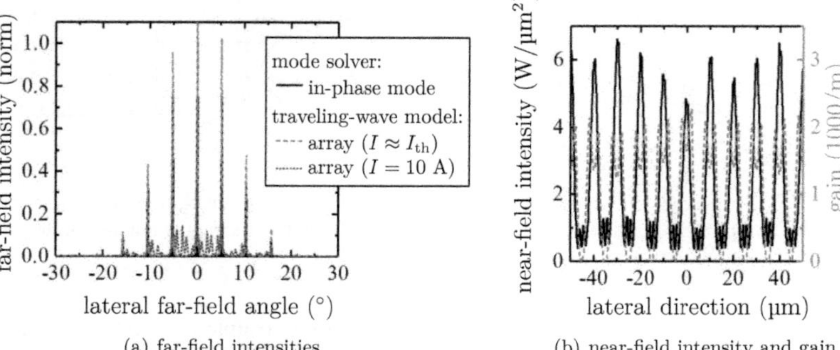

(a) far-field intensities

(b) near-field intensity and gain

Figure 7.14: (a) The far-field intensity of the in-phase mode obtained from the Helmholtz equation (6.20) (solid black) compared to the results of the traveling-wave model at an injection current near threshold (red dashed) and at a higher injection current of 10 A (blue dotted). (b) Near-field intensity and gain distribution of the traveling-wave model results at 10 A. The array stripe width and interelement distance are $s = d = 5$ µm, the effective index step between arrays elements is $\Delta n_0 = -1 \cdot 10^{-2}$, and the effective index step of the edge index-guiding trench is $\Delta n_1 = -2 \cdot 10^{-2}$ with a distance to the injection array of $L = 2.1$ µm (compare Fig. 7.13).

the array displayed in Fig. 7.14 that shows perfectly coherent emission near threshold has a single lobe power content of less than 30%. How this drawback can be overcome is discussed further below in Section 7.2.3. In the following section a mode-selection mechanism is introduced, that provides full coherence for the laser operation shown in Fig. 7.14 even at high injection currents.

7.2.2 Talbot lasers

Besides the index-guided and anti-index-guided laser arrays two additional "basic" phase locked array types are considered in literature [104], that include an longitudinal variation of injection regions and real refractive index changes. These are the Y-junction laser [115], where interference is used for mode selection, and the Talbot type spatial filter [116]. Of the two the Talbot type spatial filter is discussed here in more detail, because it forms the basis of the newly developed chessboard laser that is presented in the next section.

The Talbot-type spatial filter utilizes the properties of near-field diffraction of periodic light sources and thus is also referred to as a diffraction coupled laser array [104]. Assuming paraxial propagation, the distance at which the diffraction pattern of a periodic near-field distribution with periodicity of Λ_x repeats itself is called the Talbot length,

$$z_T = 2\Lambda_x^2 n_{\text{eff}}/\lambda_0 \equiv \Lambda_z. \tag{7.4}$$

Then at $z_T/2$ the near-field distribution of the in-phase mode repeats itself with a lateral shift of half its period, whereas the out-of-phase mode self images with a lateral shift of a fourth its period because it has twice the in-phase mode lateral period (compare Fig. 7.12(a) and (b)). For a lateral period of $\Lambda_x = 10$ µm, i.e. $s = d = 5$ µm, resonance is gained for an effective index change of $\Delta n_0 = -1 \cdot 10^{-2}$. By choosing the longitudinal

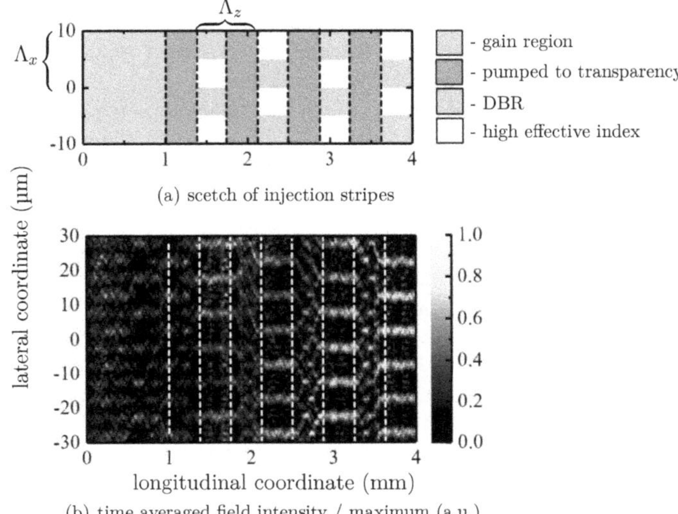

Figure 7.15: The Talbot laser. (a) Scetch of injection stripes. (b) Time averaged field intensity within the cavity divided by its maximum in every longitudinal step.

period Λ_z as the Talbot length, from Eq. (7.4) $\Lambda_z = z_T = 750$ µm is obtained. In Fig. 7.15(a) a scetch of the DBR-Talbot laser with a gain section length of $L = 4 \cdot \Lambda_z = 3$ mm and DBR section of 1 mm is shown. The studied laser has $N = 27$ periods in lateral direction and thus has an emission aperture of ≈ 270 µm. The red sections in Fig. 7.15(a) are pumped to transparency with $U = 1.38\ V$. At the grey gain regions a higher voltage ($U = 1.8$ V) is applied, so that the injection current is 15 A in total, resulting in a power of 7.5 W averaged over the last 5 ns of a 10 ns long pulse.

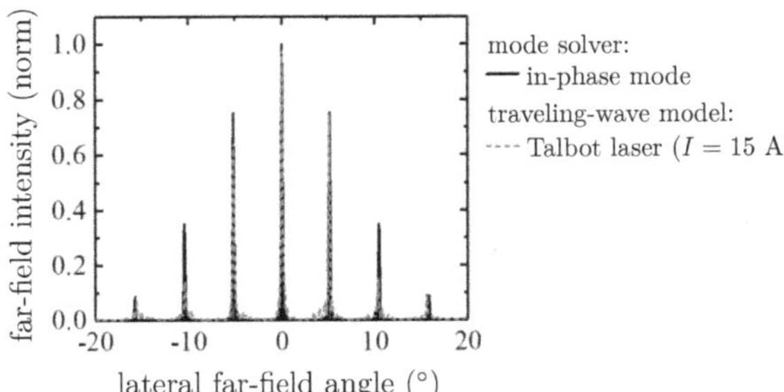

Figure 7.16: Far-field intensity distribution of a Talbot laser (red-dotted). Plotted for comparison is the in-phase mode obtained from the solution of the Helmholtz equation (black solid).

The time averaged field intensity within the cavity (averaged over the last 5 ns) divided by its maximum in every longitudinal step, Fig. 7.15(b), visualizes the expected diffraction pattern between the injection stripes.

Plotting the far-field distribution of the Talbot laser together with the far fields of the in-phase mode, clearly reveals the diffraction-limited beam operation, that is obtained at an output of 7.5 W also in the presence of spatial hole burning. However, a significant portion of the power is still radiated into the side peaks.

The simulation results show that the traveling-wave model can describe well the diffraction coupling behavior of the Talbot-type spatial filter. Furthermore important effects such as spatial hole burning that can destruct in-phase operation of anti-index-guided arrays are properly accounted for. Thus, it is well suited for the optimization of these devices and the prediction of the behavior of novel laser arrays as presented in the next section.

In [117] measurements of laser operation with a high index step of $\Delta n_0 = -3.5 \cdot 10^{-2}$ and an intra-cavity Talbot spatial filter at 15 A and 10 W output power shows a far field of 0.62° FWHM with only two side lobes, so that the central lobe power content is stated to be 60 %. The suppression of side modes can be performed by gain-loss modulation as discussed in the following section.

As mentioned early the fabrication of this anti-index-guided array has to be done using a two step epitaxy, where the high index material is selectively removed and the structure then regrown. This comparatively extensive approach can only be avoided by using index-guided structures. In the next section gain-loss modulated index-guided arrays with additional phase-tailoring are discussed to obtain a good beam quality at injection currents of up to 100 A.

7.2.3 Chessboard lasers

Longitudinal-lateral gain-loss modulation with periods corresponding to the Talbot length that fulfill Eq. (7.4), has shown to result in an anisotropic gain [109]. A scetch of parts of the injection regions in a corresponding DBR laser are shown in Fig. 7.17(a). In this so called chessboard laser beam components propagating at angles to the propagation direction, i.e. those components that result in sidelobes in the far-field intensity of arrays, Fig. 7.14(a), are attenuated, whereas the main lope is amplified. This effect has already been studied theoretically with regards to amplifiers using the traveling wave model approach [105]. However, this concept has never shown to be successful in BA lasers yet as the out-of-phase mode still has a non-negligible part in the lasing process.

In this section we will see that the out-of-phase mode can be successfully suppressed by an additional periodic modulation of the refractive index, so that extremely high brightness is predicted theoretically for the central lope under pulsed operation.

In a gain-loss material laser or amplifier the current injection path is normally tailored by implantation of the non-gain regions. In the lasers that are proposed here current-path tailoring is done by etching index-guiding trenches. Opposed to the general assumption that anti-index-guided arrays show better overall performance [104] these lasers then become index-guided and thus are much easier to fabricate. By etching index-guiding trenches not only the gain profile can be tailored but at the same time

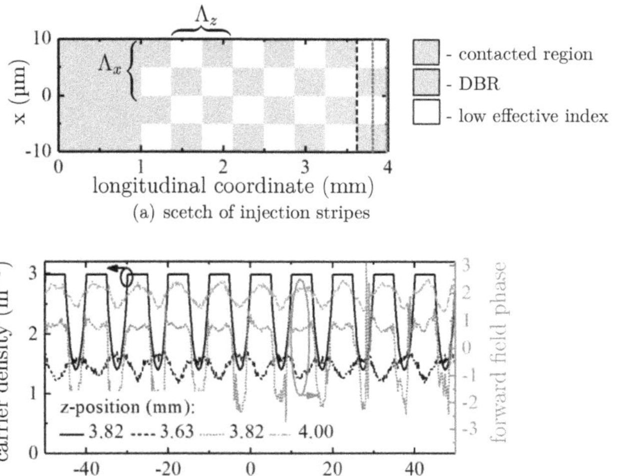

(a) scetch of injection stripes

(b) lateral carrier density distribution and forward field phase

Figure 7.17: Operation principle of a DBR-chessboard laser with lateral and longitudinal periodicity of $\Lambda_x = 10$ µm and $\Lambda_z = 750$ µ, respectively, and additional phase tailoring. (a) Scetch of injection stripes. Red dashed and dotted lines indicate the longitudinal z-positions of line plots in (b). (b) Lateral carrier density distribution (left axis - black) in the middle of the last section ($z = 3.82$ mm) and forward field phase (right axis - red) at beginning ($z = 3.63$ mm), middle ($z = 3.82$ mm) and end ($z = 4$ mm) of the last section at the last time instance after a 10 ns long simulation transient at 70 A injection current.

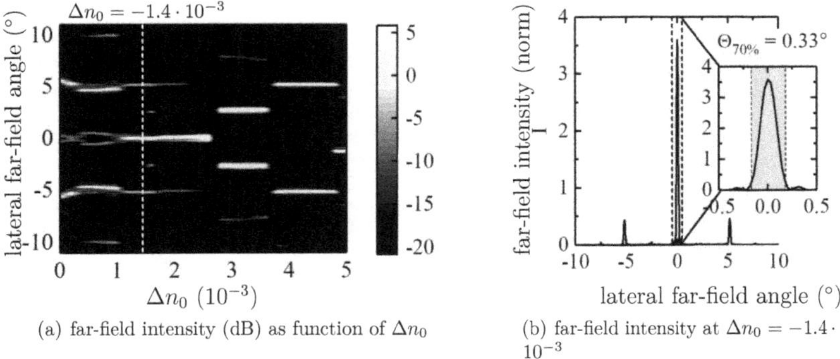

(a) far-field intensity (dB) as function of Δn_0

(b) far-field intensity at $\Delta n_0 = -1.4 \cdot 10^{-3}$

Figure 7.18: DBR-chessboard laser with $\Lambda_x = 10$ µm and $\Lambda_z = 750$ µm without spatial hole burning ($r_s = 10^{-12}$ Ω/m²). (a) Far-field intensity distribution as function of the built-in index step Δn_0. At $\Delta n_0 = -1.4 \cdot 10^{-3}$ a maximum brightness is achieved. (b) Far-field intensity for $\Delta n_0 = -1.4 \cdot 10^{-3}$. The central lope has a power content of 70% at a far-field angle of $\Theta_{70\%} = 0.33°$.

a phase shift is introduced between element and interelement areas (grey and white regions in Fig. 7.5(a)),

$$\Delta\varphi = k_0\Lambda_z/2 \cdot (n_0 - n_1). \tag{7.5}$$

In order to suppress the out-of-phase mode a $\Delta\varphi \approx \pi$ phase-shift has to be introduced between element and interelement. In this way the fields phase is tailored so that it self images with a lateral shift of half its period after propagating the distance $\Lambda_z/2$. This is illustrated in Fig. 7.17(b) for a model structure under isothermal conditions with similar parameters to structure II but with a very small series resistivity (3.46) of $r_s = 10^{-12}\ \Omega/\mathrm{m}^2$ so that spatial hole burning is not occurring. The simulation transient was 10 ns at a high injection current of 70 A. Here it is clearly visible, that light that propagates in the interelement regions will attain a π-phase shift relative to light that propagates in the injection element regions so that at the output the phase self images with a lateral shift of half its period ($z = 3.63$ mm compared to $z = 4.00$ mm).

Generally n_0 and n_1 are not constant throughout the device and depend on the carrier density and fast temperature fluctuations,

$$n_0 = -\sqrt{n'_N \cdot N} + \Delta n_{T,0} \tag{7.6}$$

$$n_1 = \Delta n_{T,1} - (\Delta n_0 + \sqrt{n'_N \cdot N}) \approx -(\Delta n_0 + \sqrt{n'_N \cdot N_{\mathrm{tr}}}), \tag{7.7}$$

where $\Delta n_{T,0}$ and $\Delta n_{T,1}$ denote the temperature increase in the high, respective low index regions. Thus the necessary built-in index variation of the trench resulting from (7.5) and $\Delta\varphi = \pi$,

$$\Delta n_0 = \lambda_0/\Lambda_z - \Delta n_{T,0} + \sqrt{n'_N \cdot N} - \sqrt{n'_N \cdot N_{\mathrm{tr}}}, \tag{7.8}$$

is difficult to ensure at every longitudinal position within the laser. Only in the absence of spatial hole burning and thermal waveguiding a nearly diffraction-limited single-lobed far field can be obtained for longitudinally and laterally constant Δn_0. In Fig. 7.18(a)

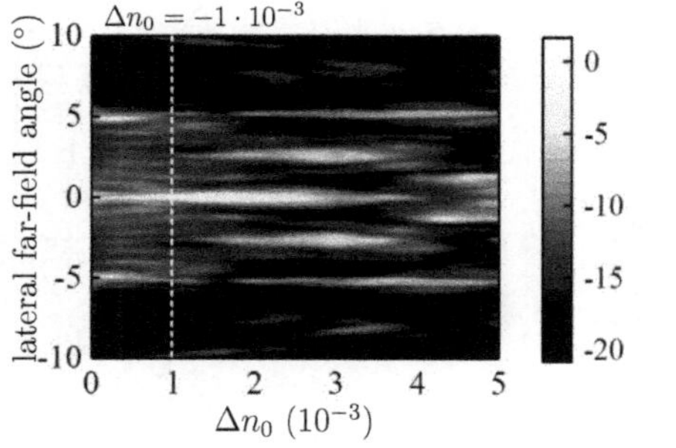

(a) far-field intensity (dB) as function of Δn_0

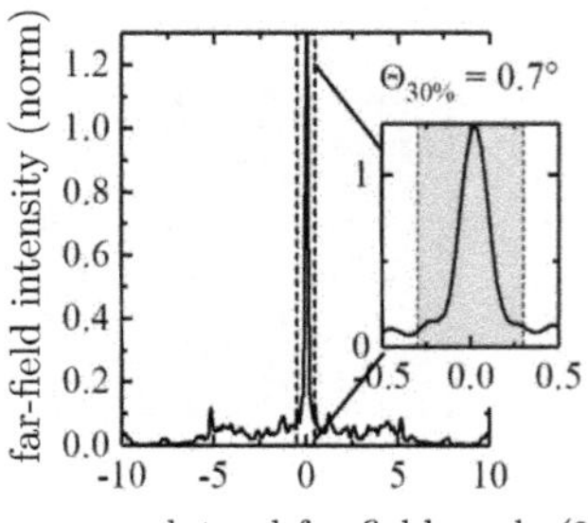

(b) far-field intensity at $\Delta n_0 = -1 \cdot 10^{-3}$

Figure 7.19: DBR-chessboard laser with $\Lambda_x = 10$ µm and $\Lambda_z = 750$ µm. (a) Far-field intensity distribution as function of the built-in index step Δn_0, at $\Delta n_0 = -1 \cdot 10^{-3}$ a maximum brightness is achieved. (b) Far-field intensity for $\Delta n_0 = -1 \cdot 10^{-3}$. The central lope has a power content of 30% at a far-field angle of $\Theta_{30\%} = 0.7°$.

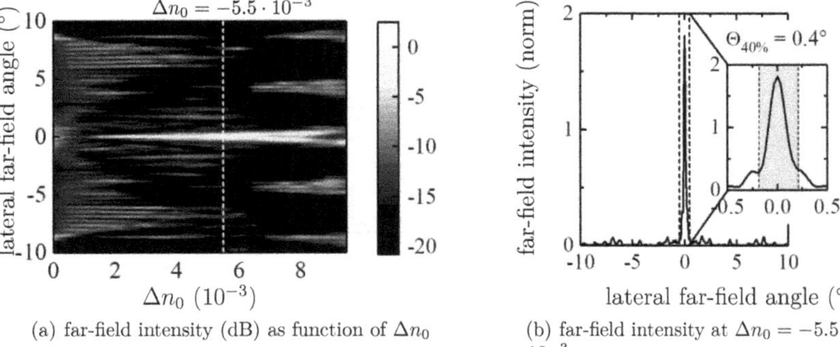

(a) far-field intensity (dB) as function of Δn_0

(b) far-field intensity at $\Delta n_0 = -5.5 \cdot 10^{-3}$

Figure 7.20: DBR-chessboard laser with $\Lambda_x = 6$ µm and $\Lambda_z = 260$ µm. (a) Far-field intensity distribution as function of the built-in index step Δn_0. At $\Delta n_0 = -5.5 \cdot 10^{-3}$ a maximum brightness is achieved. (b) Far-field intensity for $\Delta n_0 = -5.5 \cdot 10^{-3}$. The central lope has a power content of 40% at a far-field angle of $\Theta_{40\%} = 0.4°$.

the dependence of the far-field intensity on the built-in index change Δn_0 is displayed in the absence of spatial hole burning and thermal waveguiding, where the carrier density is constantly $N = 3 \cdot 10^{24}$ m^{-3} in the gain regions and $N = N_{\text{tr}} = 1.4 \cdot 10^{24}$ m^{-3} in the interelement regions, *cf.* Fig. 7.17(b). The applied voltage was kept constant at $U = 1.418$ V, resulting in a current of 70 A for $\Delta n_0 = -1.4 \cdot 10^{-3}$. In this simulation and all following simulations the built-in index step variation was only considered optically. For current confinement a constant residual layer thickness of $d_{\text{res,imp}} = 315$ nm corresponding to $\Delta n_0 = -3 \cdot 10^{-3}$ is assumed, where current spreading is already largely diminished. According to Eq. (7.8) the built-in index change at resonance is $\Delta n_0 = -1.4 \cdot 10^{-3}$ as is visible in the mapping of Fig. 7.18(a). For this case the side peaks at $\Theta = \pm\lambda/\Lambda_x = 5.2°$ are highly reduced, Fig. 7.18(b), and the far field is nearly single-lobed with a full lateral far-field angle containing 70 % of the power of $\Theta_{70\%} = 0.45°$. With the emission aperture of $w_0 = 270$ µm this yields a beam quality factor of $M^2 = \pi/\lambda_0 \cdot w_0\Theta/4 = 1.3$ close to the diffraction limit.

For the realistic structure II the field intensity distribution as function of the built-in index step Δn_0 is shown in Fig. 7.19(a). The simulation was done including fast temperature fluctuations (2.30) at an injection current of 100 A. Although the carrier density is increased close to the DBR grating to $N = 5 \cdot 10^{24}$ m^{-3}, resonance is gained at a lower Δn_0 of $-1 \cdot 10^{-3}$, Fig. 7.19(b), due to the small scale local heating below the injection stripes $\Delta n_{T,0}$ (*cf.* Eq. (7.8)). In this case the noise level is much higher compared to Fig. 7.18(b), so that the full lateral far-field angle containing 30 % of the power is $\Theta_{30\%} = 0.7°$, which results in a beam quality factor of $M^2 = 2.9$ with $w_0 = 270$ µm.

As the Δn_0 needed for resonance (7.8) is inversely proportional to the longitudinal period Λ_z, a chessboard laser with smaller Λ_z and accordingly Λ_x is less effected by variations of the build in effective index. A further advantage of smaller grating periods is a higher side mode suppression for the same gain length, because in each round trip the light simply passes through a higher number of chessboard elements. Although current spreading is already largely diminished by a trench corresponding to $\Delta n_0 = -1 \cdot 10^{-3}$

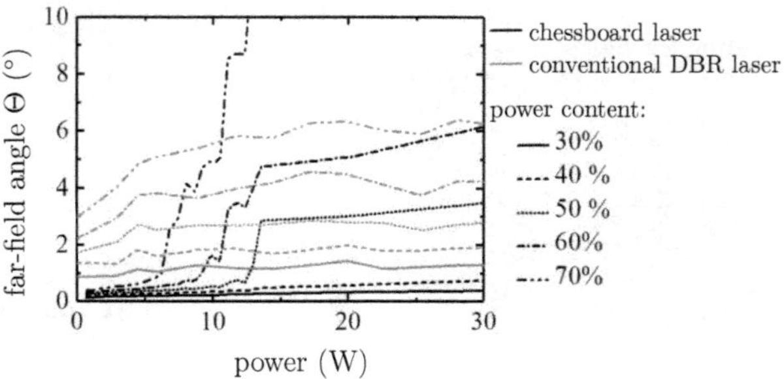

Figure 7.21: Comparison of the far-field angle of the chessboard laser with $\Lambda_z = 260$ µm, $\Lambda_x = 6$ µm and $\Delta n_0 = -5.5 \cdot 10^{-3}$ with a conventional DBR BA laser of the same length and aperture of 270 µm.

smaller grating periods demand for more deeply etched trenches and thus even better current confinement. For $\Lambda_z = 260$ µm and accordingly $\Lambda_x = 6$ µm the field intensity distribution as function of the built-in index step Δn_0 is shown in Fig. 7.20(a). As expected, the Δn_0 range for which a single-lobed far field can be obtained is wider compared to the previously discussed laser with longer periods. Furthermore, sidelobes are more effectively suppressed. The highest brightness is achieved at $\Delta n_0 = -5.5 \cdot 10^{-3}$, Fig. 7.20(b), with a far-field angle containing 40 % of the power of $\Theta_{40\%} = 0.4°$, which results in a beam quality factor of $M^2 = 1.6$ with $w_0 = 270$ µm.

The evolution of the far-field angle with varying power contents is shown in Fig. 7.21 for this chessboard laser with increasing power together with the far-field angle of a conventional DBR laser with the same length and aperture of 270 µm. At a power content of 30% the chessboard laser shows a smaller far-field angle also at high powers. When the power content is increased to at least 50% the performance of the chessboard laser is only better than that of the conventional DBR laser for output powers of approximately 10 W, because of the considerable amount of background noise that is observable in the far field.

Comparison with measurements

In Fig. 7.22 the measured and simulated far-field intensity distributions of two chessboard lasers with grating periods $\Lambda_x = 6$ and $\Lambda_x = 12$ µm and are shown for comparison at an injection current of approximately 100 A. The measurement was performed with pulses of 10 ns pulse length with a repetition rate of 10 kHz.

For dry etching an error of the build in effective index of $\sigma_{\Delta n_0} = \pm 10^{-3}$ is generally expected. The etching depth of the lasers was measured after fabrication and corresponds to a built-in index step of $\Delta n_0 = -5.5 \cdot 10^{-3}$ for the laser with grating period $\Lambda_x = 6$ and a built-in index step of $\Delta n_0 = -3 \cdot 10^{-3}$ for the laser with grating period $\Lambda_x = 12$.

In the measurement of Fig. 7.22(a) a strong central peak can be observed, which is predicted by the simulation. However, although the design parameters were chosen to result in a high central lobe power fraction, the intensity beside this central peak is significant. At $\Theta = \pm\lambda_0/(2\Lambda_x) = 4.3°$ for example distinct peaks corresponding to the out-of-phase mode are visible, but also other clear peaks emerge that may correspond to other waveguide modes. In the measurement of Fig. 7.22(b) the design parameters

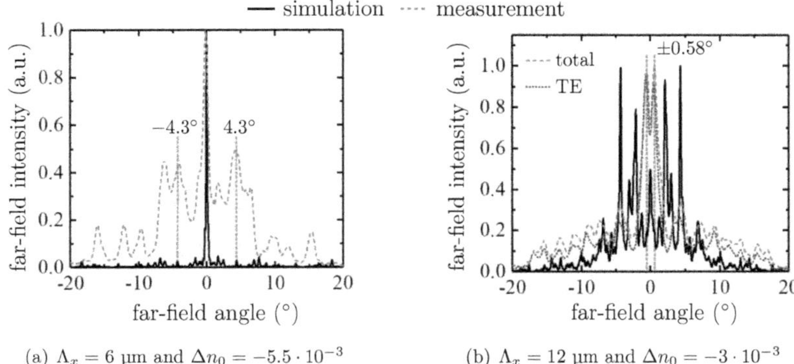

(a) $\Lambda_x = 6$ µm and $\Delta n_0 = -5.5 \cdot 10^{-3}$

(b) $\Lambda_x = 12$ µm and $\Delta n_0 = -3 \cdot 10^{-3}$

Figure 7.22: Comparison of the far-field intensity distribution of the chessboard lasers with grating periods (a) $\Lambda_x = 6$ µm and built-in index step $\Delta n_0 = -5.5 \cdot 10^{-3}$ and grating periods (b) $\Lambda_x = 12$ µm and and built-in index step $\Delta n_0 = -3 \cdot 10^{-3}$ at an injection current of approximately 100 A. In panel (b) in addition to the total field the TE polarized contribution to the field is displayed. The measurement was performed with pulses of 10 ns pulse length and repetition rate of 10 kHz to exclude heating between the pulses.*

were not at optimum and the measured far field intensity is multi-peaked with a double lobe at $\Theta = \pm 0.58°$. These double peaks correspond to the 5th waveguide mode that has a beat with period length of $\Lambda_x = \pm\lambda_0/\Theta = 90$ µm. Similar to the measurement in Fig. 7.22(a) the underground besides the central double peak is high, indicating that a variety of waveguide modes is excited.

Although simulation and experiment differ, the far field intensity can clearly be tailored by the longitudinal-lateral structuring to obtain an increased intensity around 0° far-field angle. This becomes more clearly visible in Fig. 7.23(a) where the measured profiles of the lateral far-field intensity of the same chessboard lasers with grating periods $\Lambda_x = 6$ µm and $\Lambda_x = 12$ µm are displayed together with the measured lateral far-field intensity profile of a conventional DBR BA laser of the same length and aperture of 270 µm (reference). In all cases the pulsed peak output power is approximately 35 W, which is achieved for the chessboard lasers with a pulsed peak current of 100 A and for the conventional DBR BA laser at a lower pulsed peak current of 40 A due to the greater pumped region. The profiles are normed, i.e. divided by their lateral integral, and here it is already visible that the chessboard designs result in a higher power content at very small far field angles. Furthermore the full width at half maximum (FWHM) of the central lobe is only $\Theta_{\mathrm{FWHM}} = 1°$ and $\Theta_{\mathrm{FWHM}} = 2.8°$ for the chessboard lasers with grating periods $\Lambda_x = 6$ µm and $\Lambda_x = 12$ µm, respectively, and $\Theta_{\mathrm{FWHM}} = 14.9°$ for the conventional DBR BA laser. At a power content of less than 35% the far field angle of the chessboard laser with grating period $\Lambda_x = 12$ µm becomes smaller compared to the reference, Fig. 7.23(b), whereas the far field angle containing 95% of the power $\Theta_{95\%}$ is approximately twice as high.

While the measurements indicate that this approach points in the right direction, the optimum operation point has not been reached yet and there still remains a discrepancy between measurements and simulation. The distribution of the effective index which

* The author thanks H. Christopher, J. Hopp and A. Klehr for the pulsed measurements.

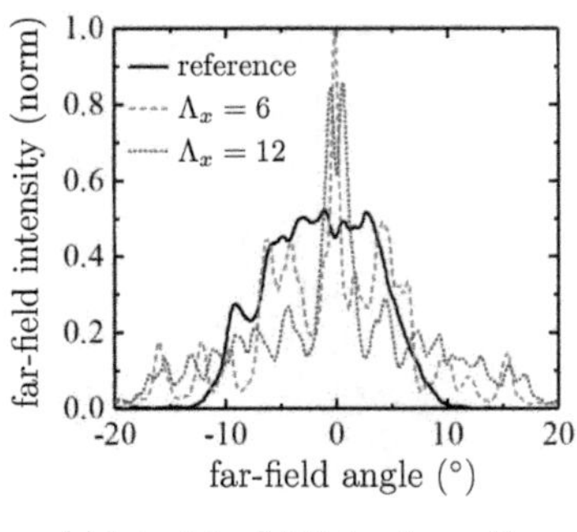

(a) lateral far-field intensity profile

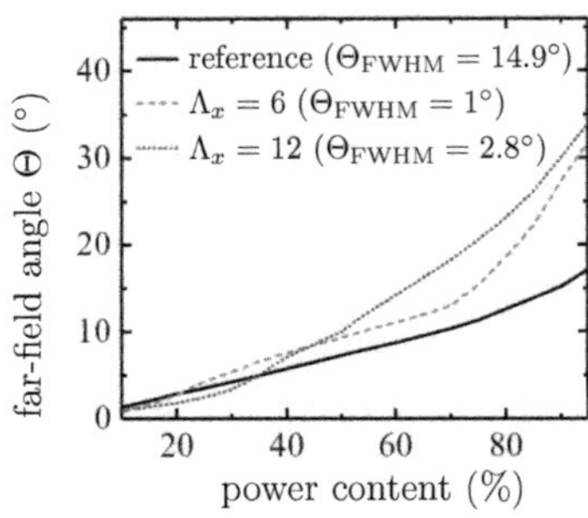

(b) far-field angle as function of power content

Figure 7.23: Comparison of the measured (a) lateral far-field intensity profile and (b) far-field angle containing different power fractions of the chessboard lasers with a conventional DBR BA laser of the same length and aperture of 270 µm (reference) at an output power of $P_{\text{out}} = 35$ W. The structured lasers with grating periods $\Lambda_x = 6$ µm and $\Lambda_x = 12$ µm have a built-in index step of $\Delta n_0 = -5.5 \cdot 10^{-3}$ and $\Delta n_0 = -3 \cdot 10^{-3}$, respectively. The structured lasers are driven at a pulsed peak injection current of approximately 100 A and the reference laser with approximately 40 A. The measurement was performed with pulses of 10 ns pulse length and repetition rate of 10 kHz to exclude heating between the pulses.*

is the basis of the calculation of Δn_0 in Eq. (7.8) might be incorrectly predicted, for example when the temperature increase in the high index regions where the guided mode is localized is actually higher. Then the design parameter Δn_0 becomes smaller than calculated here. It shall be noted that the measured far field in Fig. 7.22(a) actually resembles more the mapping of the far fields of Fig. 7.20 at higher values for Δn_0, which might be an indication of this effect.

A further possibility is that the longitudinal variation of carrier density and temperature is higher than expected, which would lead to a higher background because the resonance condition (7.5) is not met in more parts of the resonator. Similarly, fabricated built-in index-guiding trenches are not rectangular but V-shaped and have a minimum etching depth from which the experimental Δn_0 is calculated. Furthermore the not rectangular pulse shape of the current source including rise and decay time can lead to a different distribution of the carriers in the resonator.

In this work the optical field is represented by its transverse-electric x-component, see Eq. (2.1). Shear strain, which can be induced by build in stress, is not contained in the model and can cause a rotation of the polarization axis of the optical field, so that an electric field vector that initially oscillated in the active region plane will be rotated according to the permittivity tensor of the material [108]. In Fig. 7.22(b) two measurements of the far-field profile are displayed: Firstly a far-field measurement containing the total field intensity drawn as red dashed line and secondly the field intensity that is obtained by inserting a polarization filter into the optical path so that only light passes where the electric field component oscillates in the active region plane as dark red dotted line, termed here "TE". These measurements show that approximately 80 % of the measured light field is TE polarized. Although this contribution has a

* The author thanks H. Christopher, J. Hopp and A. Klehr for the pulsed measurements.

better side suppression compared to the total field intensity, the far field still shows a tremendous background, so that only little improvement for the agreement of simulation and measurement is to be expected by an inclusion of shear strain.

7.2.4 Conclusions

In this section coherently coupled laser arrays and variations are discussed for pulsed operation. Results found in literature can be well reproduced by the traveling wave model. Furthermore a new variation of the array laser, the "chessboard laser", is presented for which a narrow single-lobed far field is predicted also at high pulsed currents of 100 A.

If the resonance condition is met, anti-index-guided in-phase coupled arrays show a narrow single lobe around 0° far-field angle at currents near threshold. For higher currents noise is added as a result of spatial hole burning. A selection mechanisms for the in-phase mode can be a Talbot spatial filter, i.e. alternate inclusion of free running sections that are pumped to transparency and have a length that corresponds to half the Talbot length. When this is done the laser shows stable operation also far above threshold. However still, a significant proportion of the light is transmitted to higher far-field angles.

A new type of laser is gained by an additional longitudinal structuring of the injection stripe of an index-guided laser array ("chessboard laser"), which results in anisotropic gain. By additional phase tailoring using etched index-guiding trenches the out-of-phase mode can be successfully suppressed and a single-lobed far field with a far-field angle containing 40% of the power of $\Theta_{40\%} = 0.4°$ is obtained theoretically at an injection current of 100 A under pulsed operation.

While the measurements indicate that this approach points in the right direction, the optimum operation point has not been reached yet and there still remains a discrepancy between measurements and simulation. The discrepancy may originate from an altered distribution of the effective index for example by a higher temperature increase in high index regions, a stronger longitudinal variation of carrier density and temperature or not rectangular shaped trenches or injection current pulses. Strain effects, that are not included in the simulation model, are found to increase the background of the far-field intensity at higher far-field angles.

Chapter 8

Summary and outlook

For the application of broad-area (BA) lasers a high output power P_{out} and good beam quality, i.e. small beam parameter product BPP_{lat}, that are both combined in the target figure brightness $B_{\text{lat}} = P_{\text{out}}/\text{BPP}_{\text{lat}}$ is of great importance. To understand the underlying spatio-temporal phenomena and to apply this knowledge in order to reduce costs for brightness optimization, a self-consistent simulation tool taking into account all essential processes is vital. Firstly, in this work a quasi-three-dimensional opto-electronic and thermal model is presented, that describes well essential qualitative characteristics of real devices and in particular the inherently non-stationary behavior of BA lasers. Secondly, the origins of brightness degradation, i.e. power saturation and the modulated and not diffraction-limited field profile are investigated. And lastly, designs that mitigate those effects that limit the lateral brightness B_{lat} under pulsed and continuous-wave (CW) operation are discussed.

Quasi-three-dimensional opto-electronic and thermal model

The challenge of the very different time and length scales involved in the lasing process is met by using different simulation domains and an iterative approach to self-consistently couple the stationary heat-flow equation to the dynamic electro-optical equations, thus enabling the efficient simulation of real devices. The presented model is based on an existing optical field model, for which the traveling-wave equations in the lateral and longitudinal (direction of propagation) (x, z)-plane are solved together with a lateral diffusion equation of the excess carriers in the same plane. In this work the model is extended by an advanced description of the injection current density. In this way current spreading in the p-doped layers and current self-distribution, to properly account for spatial hole burning effects, are included. Furthermore a temperature model is presented that takes into account refractive index changes due to short-time local heating near the active region, which is important for the study of short pulse operation. Additionally the formation of a stationary temperature profile under CW operation can be incorporated by iteratively calculating time-averaged heat sources from transient calculations of the opto-electronic model and utilizing them to solve the static heat-flow equation. The resulting thermally induced effective index profile has a significant influence on the wave propagation and the temperature dependence of parameters results in thermal power saturation.

Although in this work, the simulation model is only used for the investigation and improvement of BA lasers, it is also suited for all kinds of active narrow-banded device simulations, such as amplifiers or ridge-waveguide lasers including integrated optical components [118].

Origins of lateral brightness degradation

A second major part of this thesis is dedicated to the reasons of brightness degradation in BA lasers, either as a result of power saturation or due to a declining beam quality. Here, particular attention is payed to the origins of the multi-peaked and not diffraction-limited field profile.

Under CW operation power saturation is mainly attributed to device heating as loss mechanism such as recombination, free carrier absorption and leakage currents are enhanced, whereas under pulsed operation, when nanosecond-current pulses with low repetition rates are applied, thermally induced power rollover is believed to be negligible. Still, BA lasers experimentally exhibit a strong power saturation, which is identified to be partly caused by spatial hole burning, current spreading and two-photon absorption. It is revealed that spatio-temporal power variations play a role in the power saturation process and should not be neglected. Further effects that could influence the power-current characteristics are free carrier absorption by two-photon absorption generated carriers (indirect two-photon absorption) as discussed in Refs. [11, 10] and capture-escape processes as discussed in Ref. [81], which should be considered for further model expansions.

In this thesis two mechanisms that are generally referred to when it comes to the understanding of the multi-peaked lateral field profile are discussed. The Bespalov Talanov modulation instability [47] describes the spontaneous break-up of the optical field into small filaments as a result of a focusing Kerr non-linearity. In semiconductor lasers an indirect Kerr-type non-linearity results from spatial hole burning, because the refractive index in areas of high intensity rises, thus creating a local waveguide. For a hypothetical steady state an equation for this indirect Kerr-type nonlinearity can be formulated [27]. In BA lasers the interaction of forward- and backward-propagating fields and their non-stationarity have to be considered. Furthermore in the presented theory gain changes are neglected, which have a defocusing effect, because of gain reduction in areas of high field intensity. Thus, the traveling-wave simulations could not support the theory of filamentation by this indirect Kerr-type non-linearity.

An alternative understanding of the transverse instabilities and multi-peaked field structure of BA lasers bases on the simultaneous lasing of a large number of lateral waveguide modes [88, 89, 16]. In the traveling-wave approach no pre-assumptions regarding modes is made, however, any field can be expressed as a linear combination of a complete set of eigenmodes. Here in a post-processing the optical field obtained by a laser simulation with the traveling-wave model is decomposed into lateral waveguide modes. The results indicate that in lasers with a lateral waveguide a clear mode structure is visible which is neither destroyed by the dynamics nor by longitudinal effects.

With increasing current a broadening of the near and far fields can be observed, which is commonly referred to as near or far field "blooming". In the absence of a thermally induced waveguide transverse instabilities arise due to lateral spatial hole burning and the nonlinear interaction between the electromagnetic field and the gain material. It turns out that the degradation of the beam quality can be well understood in the mode picture by the suppression of higher order lateral modes. However, the mode picture has also its limitations when it comes to anti-index-guided structures.

In the past thermal far-field blooming under CW operation was theoretically treated usind stationary temperature distributions [18, 19, 20, 17]. In this work, the application of the time-dependent approach to CW operation yields strong spatio-temporal sub-nanosecond fluctuations of the heat sources. However, they have a negligible influence on the wave propagation, whereas the effects of the time-averaged longitudinally varying temperature profile (that arises in lasers with assymetric facet reflectivity values) confirm those obtained with stationary temperatures in a previous publication [17]. The impact of this profile is significant and leads to a near-field shrinkage at the anti-reflection coated facet. An unfavorable carrier accumulation at the stripe edges and an enhanced power density that lowers the threshold for facet damage is the result.

Thermal waveguiding is usually neglected under pulsed operation, because thermal build-up times up to milliseconds are much longer than the pulse lengths. However, the heat is generated near the active layer in the same region where the guided wave is localized. This region is small and its thermal build-up time is much shorter than that of the whole device. The simulation of an exemplary laser operation with 10 ns long pulses reveals a fast-growing thermally induced waveguide initializing the known stationary "thermal lense". Although the temperature increase is small, a waveguide that is introduced within a few-nanoseconds-long pulse can result in a transition from a gain-guided to an index-guided structure, leading to a near-field narrowing.

Designs for lateral brightness improvement

The final part of this thesis concerns itself with a lateral or longitudinal-lateral structuring of the laser device to alter the complex refractive index in order to improve the brightness. It is shown, that an implantation of the contact layer next to the injection stripe to counteract current spreading or a lowered p-layer resistivity to reduced spatial hole burning are generally favorable.

The narrowed near-field intensity resulting from the longitudinally varying temperature distribution under CW operation can be additionally tailored by appropriate lateral index-guiding trenches. On the one hand, they result in a beam quality degradation as higher order lateral modes have a higher contribution to the lasing process. On the other hand, if the laser threshold for facet damage and its efficiency are taken into account, etching index-guiding trenches has a positive effect, which is supported by measurements. An optimum trade-off between efficiency and beam quality could be found by varying trench parameters such as the effective index step, trench width and distance of the trench to the injection stripe.

Filtering modes by adequate lateral-longitudinal structuring of the contact region in order to improve beam quality have been proposed in numerous publications [103, 104, 105, 106, 107]. In this work a novel laser design is presented, that bases on the findings of [109], which shows that longitudinal-lateral gain-loss modulation with periods corresponding to the Talbot length result in an anisotropic gain. By additional phase tailoring the preferred mode is excited and a single-lobed far field with a far-field angle containing 40% of the power of $\Theta_{40\%} = 0.4°$ is obtained theoretically at an injection current of 100 A under pulsed operation. While the measurements indicate that this approach points in the right direction, here the optimum operation point has not been reached yet. Furthermore, there still remains a discrepancy between measurements and simulation which may originate either from the longitudinal refractive index variation or the incorrectly predicted magnitude of the refractive index change.

Outlook

Several improvements should be considered for further model expansions.

Comparing temperatures of the active zone calculated by the bulk heat-flow equations with measured ones, indications of an inhibition of heat flow can be found that possibly arise due to thermal boundary resistance of hetero interfaces. For further improving the simulation model regarding pulsed applications with pulses longer than few nanoseconds, it is therefore desirable to solve the time-dependent heat-flow equation within a small vicinity of the active region of ≈ 1 µm in vertical direction taking into account this boundary effect.

Under CW operation the stationary heat-flow equation is solved by neglecting heat flow in longitudinal (propagation) direction, because the characteristic scales in this direction are much larger than the scales of the transverse (x, y)-plane. For the simulation of longitudinally noninvariant designs under CW operation, such as tapered [106] or chessboard [105] designs longitudinal heat flow should be included.

Shear strain induced by build in stress causes a rotation of the polarization axis of the optical field, so that an electric field vector that initially oscillated in the active region plane, will be rotated according to the permittivity tensor of the material [108]. Furthermore, strain in the quantum well active region can change the gain coefficients of the light field for which the electric field vector oscillates in the active region plane ("TE-polarized") or perpendicular to it ("TM-polarized"). Accordingly, the model can be improved by considering an anisotropic complex tensor for the permittivity. Then equations for the TE and TM polarized light have to be coupled to each other via a coupling coefficient and both to the carrier density reservoir with the different gain coefficients g_{TE} and g_{TM}. However, in this theses measurement examples with a polarization filter show that the electric field vector of the emitted light field mainly oscillates in the active region plane. Accordingly the inclusion of shear strain is not expected to significantly improve the agreement of simulation and measurement.

Although the overall trend of operating figures, such as lateral near-field width, far-field angle and brightness, as function of current is correctly predicted, a significant discrepancy between measurements and simulation is, that the simulated far-field angle is much smaller than measured. A higher modulation of the lateral carrier density distributions and the resulting interaction with the optical field via the carrier density induced refractive index could lead to a broader lateral far field [108]. However, also for a constant injection current density a lateral carrier density modulation of the necessary magnitude does not arise using the time-dependent traveling wave model. An interesting experimental approach that gives insight into the carrier density variation within the active region is published in [119]. Here longitudinal spatial hole burning could be quantified by measuring the spontaneous emission. Similarly, clarity on the actual lateral carrier density modulation in the active region could be gained, which, however, is much more challenging because a much better spatial resolution is necessary.

Appendix A

Simulation parameters

In this section the determination of simulation parameters will be discussed and the simulation parameters of the utilized structures are given.

The parameters describing the dependence of optical gain (2.31) and refractive index (2.29) on the carrier density, i.e. the differential modal effective index n'_N, the differential modal gain g', gain clamping carrier density N_0 and transparency carrier density $N_{\rm tr}$, as well as the parameters describing gain dispersion (2.40), used in the WIAS-BALaser model are obtained from an adaption to a microscopic gain model [56]. It takes into account all inter and intraband transitions between the lowest conduction band and the three uppermost non-parabolic valence bands. The dependence of gain and carrier induced effective index change on the carrier density are exemplary shown in Fig. A.1 for structure II (Tab. A.3).

α_0 is the absorption coefficient calculated from the cross-sections of free carrier absorption f_n and f_p for electrons and holes, respectively and the electron n and hole p densities within the vertical structure,

$$\alpha_0 = f_n n + f_p p \quad \text{for} \quad y \notin \text{active region}. \tag{A.1}$$

The electron n and hole p densities are determined by assuming charge neutrality in all layers (3.5) so that they are given by the doping profile. The cross section for free

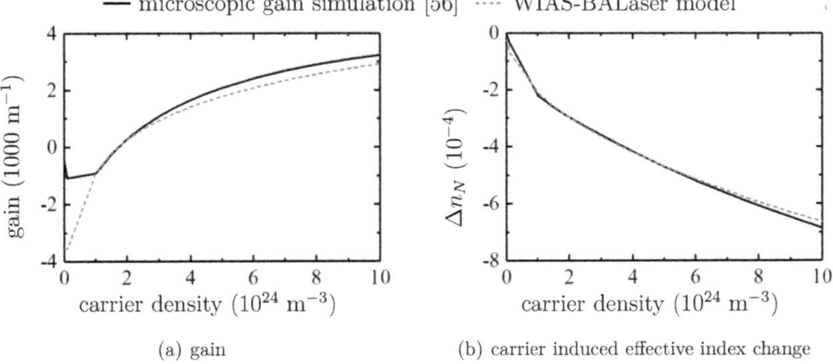

(a) gain

(b) carrier induced effective index change

Figure A.1: Exemplary adaption of gain and carrier induced effective index change Δn_N of the model parameters to the results of a microscopic gain simulation [56] for structure II (Tab. A.3) at the central wavelength.

carrier absorption is $f_n = 4 \cdot 10^{-22}$ m^2 for n-doped layers and $f_p = 12 \cdot 10^{-22}$ m^2 for p-doped layers and $f_N = 16 \cdot 10^{-22}$ m^2 in the active region [120].

The effective coefficients for internal background absorption $\alpha_{0,\mathrm{eff}}$, two-photon absorption β' and the optical Kerr coefficient n_2' are calculated according to Eqs. (2.33), (2.36) and (2.37) respectively, assuming step-wise constant local coefficients $\alpha_{0,i}$, β_i and $n_{2,i}$ in the ith-layer.

Other parameters such as B, C, the gain compression factor ϵ_s are taken from literature given in the parameters lists. If not denoted otherwise the simulation parameters given in Tab. A.1 to A.7 are used.

Temperature dependence of model parameters

$T_0 = 300$ K is the reference temperature for which the parameters have been calculated. Conveniently it is set equal to the heat sink temperature T_{HS} occurring in the boundary condition for the heat transport equation. In what follows T_{a} is the temperature in the active layer as introduced before.

differential gain

$$g'(T_{\mathrm{a}}) = g'(T_0) + \partial_T g'(T_{\mathrm{a}} - T_0) \tag{A.2}$$

transparency carrier density

$$N_{\mathrm{tr}}(T_{\mathrm{a}}) = N_{\mathrm{tr}}(T_0) + \partial_T N_{\mathrm{tr}}(T_{\mathrm{a}} - T_0) \tag{A.3}$$

carrier density induced effective index

$$n_N'(T_{\mathrm{a}}) = n_N'(T_0) + \partial_T n_N'(T_{\mathrm{a}} - T_0) \tag{A.4}$$

free carrier absorption

$$f_N(T_{\mathrm{a}}) = f_N(T_0) \left(\frac{T_{\mathrm{a}}}{T_0}\right)^{\gamma_{f_N}} \exp\left[\frac{E_{\mathrm{a},f_N}}{k_B T_0}\left(1 - \frac{T_0}{T_{\mathrm{a}}}\right)\right] \quad \text{from [5]} \tag{A.5}$$

density of states

$$N_{\mathrm{c}}(T_{\mathrm{a}}) = N_{\mathrm{c}}(T_0) \left(\frac{T_{\mathrm{a}}}{T_0}\right)^{\gamma_{N_{\mathrm{c}}}} \tag{A.6}$$

$$N_{\mathrm{v}}(T_{\mathrm{a}}) = N_{\mathrm{v}}(T_0) \left(\frac{T_{\mathrm{a}}}{T_0}\right)^{\gamma_{N_{\mathrm{v}}}} \tag{A.7}$$

hole mobility in active region

$$\mu_p(T_{\mathrm{a}}) = \mu_p(T_0) \left(\frac{T_{\mathrm{a}}}{T_0}\right)^{\gamma_{\mu_p}} \quad \text{from [60]} \tag{A.8}$$

recombination parameters

$$A(T_{\mathrm{a}}) = \frac{1}{\tau_{\mathrm{SRH}}(T_{\mathrm{a}})} = \frac{1}{\tau_{\mathrm{SRH}}(T_0)} \left(\frac{T_{\mathrm{a}}}{T_0}\right)^{\gamma_A} \tag{A.9}$$

$$B(T_{\mathrm{a}}) = B(T_0) \left(\frac{T_{\mathrm{a}}}{T_0}\right)^{\gamma_B} \quad \text{from [121, 5]} \tag{A.10}$$

$$C(T_{\mathrm{a}}) = C(T_0) \left(\frac{k_{\mathrm{B}} T_{\mathrm{a}}}{E_{\mathrm{a,C}}}\right)^{\gamma_C} \exp\left[\frac{E_{\mathrm{a},C}}{k_B T_0}\left(1 - \frac{T_0}{T_{\mathrm{a}}}\right)\right] \quad \text{from [45, 121]} \tag{A.11}$$

detuning of peak of material gain Lorentzian

$$\delta\omega(T_{\mathrm{a}}) = \delta\omega(T_0) + \partial_T \delta\omega(T_{\mathrm{a}} - T_{\mathrm{ref}}) \tag{A.12}$$

Table A.1: Simulation parameters for structure I

symbol	parameter	value	unit	reference
n_{g}	group refractive index	3.8		
$\bar{n}$	reference refractive index	3.3		
λ	central wavelength	$905 \cdot 10^{-9}$	m	
r_0	rear facet field amplitude reflectivity	$\sqrt{0.6}$		
r_L	front facet field amplitude reflectivity	$\sqrt{0.01}$		
d	thickness of active region	$12 \cdot 10^{-9}$	m	
Γ	confinement factor in active region	$13 \cdot 10^{-3}$		
L_{DBR}	length of DBR section	$1 \cdot 10^{-3}$	m	
κ	DBR coupling coefficient			
	for $(x, z) \in$ DBR	2000	m^{-1}	
	for $(x, z) \notin$ DBR	0	m^{-1}	
g'	differential modal gain	2100	m^{-1}	1,4
N_{tr}	transparency carrier density	$1.4 \cdot 10^{24}$	m^{-3}	1
N_0	gain clamping carrier density	$2 \cdot 10^{23}$	m^{-3}	1
ϵ_{s}	gain compression factor	$13 \cdot 10^{-26}$	m^3	[49] 4
g_{r}	amplitude of Lorentzian	4500	m^{-1}	1
γ	Lorentzian half width at half maximum	$1.5 \cdot 10^{13}$	s^{-1}	1
$\delta\omega$	detuning of Lorentzian	0	s^{-1}	1
$\alpha_{0,\mathrm{eff}}$	effective internal background absorption	55	m^{-1}	2
f_{N}	cross section for free carrier absorption	$16 \cdot 10^{-22}$	m^2	2,3
β'	effective two-photon absorption coefficient	$8.7 \cdot 10^{-5}$	W^{-1}	
n'_N	carrier density induced differential index	$2 \cdot 10^{-31}$	m^3	1
n'_2	effective Kerr coefficient	$-4.6 \cdot 10^{-13}$	mW^{-1}	
E_{g}	band gap energy	1.37	eV	
N_{c}	conduction band density of states	$1 \cdot 10^{24}$	m^{-3}	
N_{v}	valence band density of states	$5 \cdot 10^{24}$	m^{-3}	
n_0	equilibrium electron carrier density	$1 \cdot 10^{22}$	m^{-3}	
p_0	equilibrium hole carrier density	$3.6 \cdot 10^{4}$	m^{-3}	
μ_p	hole mobility in active region	$300 \cdot 10^{-4}$	$\mathrm{m}^2(\mathrm{Vs})^{-1}$	
A	Shockley Read Hall recomb. coefficient	$3.3 \cdot 10^{8}$	s^{-1}	3
B	spontaneous emission recomb. coefficient	$2.2 \cdot 10^{-16}$	$\mathrm{m}^3\mathrm{s}^{-1}$	
C	Auger recomb. coefficient	$4 \cdot 10^{-42}$	$\mathrm{m}^6\mathrm{s}^{-1}$	

1 obtained from a fit to results of a microscopic gain model [56]

2 free carrier absorption cross sections [120]: $f_n = 4 \cdot 10^{-22}$ m^2 (n-doped layers), $f_p = 12 \cdot 10^{-22}$ m^2 (p-doped layers), $f_n + f_p$ (active region)

3 validated by fit to experimental data

4 includes active region confinement factor

Table A.2: Simulation parameters for electrical model (structure I)

	layer thickness d_i (nm)	electrical conductivity σ_i σ_i (Ω^{-1}m^{-1})
contact	100	100*
	60	2320
cladding	530	1150
	100	880
	100	500
confienment	100	240
	430	145
	100	85

* includes a contact resistance of $r = 10^{-9}$ Ωm^2

Table A.3: Simulation parameters for structure II

symbol	parameter	value	unit	reference
n_{g}	group refractive index	3.9		
$\bar{n}$	reference refractive index	3.4		
λ	central wavelength	$910 \cdot 10^{-9}$	m	
r_0	rear facet field amplitude reflectivity	$\sqrt{0.95}$		
r_L	front facet field amplitude reflectivity	$\sqrt{0.01}$		
d	thickness of active region	$7 \cdot 10^{-9}$	m	
Γ	confinement factor in active region	$6.6 \cdot 10^{-3}$		
κ	DBR coupling coefficient	0	m^{-1}	
g'	differential modal gain	1655	m^{-1}	1,4
N_{tr}	transparency carrier density	$1.7 \cdot 10^{24}$	m^{-3}	1
N_0	gain clamping carrier density	$2 \cdot 10^{23}$	m^{-3}	1
ϵ_{s}	gain compression factor	$6.6 \cdot 10^{-26}$	m^3	[49][4]
g_{r}	amplitude of Lorentzian	6000	m^{-1}	1
γ	Lorentzian half width at half maximum	$2 \cdot 10^{13}$	s^{-1}	1
$\delta\omega$	detuning of Lorentzian	0	s^{-1}	1
$\alpha_{0,\mathrm{eff}}$	effective internal background absorption	47	m^{-1}	2
f_{N}	cross section for free carrier absorption	$16 \cdot 10^{-22}$	m^2	2
β'	effective two-photon absorption coefficient	10^{-4}	W^{-1}	
n'_N	carrier density induced differential index	$4.4 \cdot 10^{-32}$	m^3	1
n'_T	temperature induced differential index	$2.5 \cdot 10^{-4}$	K^{-1}	[122]
n'_2	effective Kerr coefficient	$-5.5 \cdot 10^{-12}$	mW^{-1}	
E_{g}	band gap energy	1.36	eV	
N_{c}	conduction band density of states	$1 \cdot 10^{24}$	m^{-3}	
N_{v}	valence band density of states	$5 \cdot 10^{24}$	m^{-3}	
n_0	equilibrium electron carrier density	$1 \cdot 10^{22}$	m^{-3}	
p_0	equilibrium hole carrier density	$3.6 \cdot 10^{5}$	m^{-3}	
μ_p	hole mobility in active region	$330 \cdot 10^{-4}$	$\mathrm{m}^2(\mathrm{Vs})^{-1}$	
A	Shockley Read Hall recomb. coefficient	$5 \cdot 10^{8}$	s^{-1}	
B	spontaneous emission recomb. coefficient	$1 \cdot 10^{-16}$	$\mathrm{m}^3\mathrm{s}^{-1}$	[60]
C	Auger recomb. coefficient	$4 \cdot 10^{-42}$	$\mathrm{m}^6\mathrm{s}^{-1}$	[60]
r_{th}	thermal transmission resistance	$8.3 \cdot 10^{-6}$	$\mathrm{Km}^2\mathrm{W}^{-1}$	

1 obtained from a fit to results of a microscopic gain model [56]

2 free carrier absorption cross sections [120]: $f_n = 4 \cdot 10^{-22}$ m^2 (n-doped layers), $f_p = 12 \cdot 10^{-22}$ m^2 (p-doped layers), $f_n + f_p$ (active region)

4 includes active region confinement factor

Table A.4: List of step-wise constant simulation parameters for the electrical and thermal model of the ith layer (structure II). d_i: layer thickness. $\kappa_{L,i}$: thermal conductivity. $c_{h,i}$: layer thermal capacity. σ_i: layer electrical conductivity. $n_{r,i}$: layer refractive index. α_i: layer absorption coefficient. β_i: layer two-photon absorption coefficient.

symbol	d_i	$\kappa_{L,i}$	$c_{h,i}$	σ_i	$n_{r,i}$	$\alpha_{0,i}$	β_i
unit	µm	$\mathrm{W(Km)^{-1}}$	$\mathrm{JK^{-1}m^{-3}}$	$\mathrm{(Wm)^{-1}}$		$\mathrm{m^{-1}}$	$\mathrm{cmGW^{-1}}$
submount	250.000	200.0	$2.82 \cdot 10^6$	0.0	0.00	0	0.0
gold contact	9.000	70.0	$2.47 \cdot 10^6$	0.0	0.00	0	0.0
contact	0.100	44.0	$1.74 \cdot 10^6$	99.6*	3.58	24000	0.0
	0.700	44.0	$1.74 \cdot 10^6$	2780.0	3.58	1200	0.0
p-cladding	0.060	11.5	$1.72 \cdot 10^6$	1950.0	3.33	2400	12.9
	0.600	12.9	$1.71 \cdot 10^6$	1120.0	3.13	2400	4.7
p-confine-	0.033	12.9	$1.71 \cdot 10^6$	664.0	3.13	1200	4.7
ment	0.033	10.9	$1.72 \cdot 10^6$	492.0	3.24	660	9.8
	0.083	12.2	$1.73 \cdot 10^6$	196.0	3.36	120	13.9
	0.050	13.2	$1.73 \cdot 10^6$	135.0	3.40	66	14.8
	0.050	14.6	$1.73 \cdot 10^6$	34.4	3.43	12	15.7
QW	0.007	10.5	$1.76 \cdot 10^6$	0.0	3.65	4	0.0
n-confine-	0.040	13.2	$1.73 \cdot 10^6$	0.0	3.40	24	14.8
ment	1.750	12.0	$1.72 \cdot 10^6$	0.0	3.35	45	13.5
	0.250	11.7	$1.72 \cdot 10^6$	0.0	3.35	280	13.3
	0.250	11.6	$1.72 \cdot 10^6$	0.0	3.34	560	13.1
n-cladding	1.750	11.5	$1.72 \cdot 10^6$	0.0	3.33	800	12.9
	0.100	15.9	$1.73 \cdot 10^6$	0.0	3.45	801	16.3
substrate	130.000	44.0	$1.74 \cdot 10^6$	0.0	0.00	0	0.0
insulation	d_{res}	5.0	$1.76 \cdot 10^6$	0.0	0.00	0	0.0

* includes a contact resistance of $r = 10^{-9}\ \Omega\mathrm{m}^2$

Table A.5: Adapted parameters for structure II.2 and temperature dependence

symbol	parameter	value	unit	reference
$\alpha_{0,\text{eff}}$	effective internal background absorption	40	m^{-1}	3
A	Shockley Read Hall recomb. coefficient	$2 \cdot 10^{8}$	s^{-1}	3
r_{th}	thermal transmission resistance	$5.5 \cdot 10^{-6}$	Km^2W^{-1}	3
$\partial_T n'_N$	change of carrier-dependent index-change with temperature	$-8 \cdot 10^{-36}$	m^3K^{-1}	1
$\partial_T g'$	change of differential modal gain with temperature	10	$(\text{mK})^{-1}$	1
$\partial_T N_{\text{tr}}$	change of transparency carrier density with temperature	$2 \cdot 10^{22}$	$\text{m}^{-3}\text{K}^{-1}$	1
E_{a,f_N}	activation energy (cross section for free carrier absorption)	0	eV	[5]
γ_{f_N}	exponent temperature change (cross section for free carrier absorption)	1.5		[5]
γ_{N_c}	exponent temperature change (conduction band density of states)	1		[60]
γ_{N_v}	exponent temperature change (valence band density of states)	3/2		[60]
γ_{μ_p}	exponent temperature change (hole mobility)	-2		[60]
γ_A	exponent temperature change (Shockley Read Hall recombination coefficient)	0		[5]
γ_B	exponent temperature change (spontaneous emission recombination coefficient)	-1		[121, 5]
$E_{\text{a},C}$	activation energy (Auger recombination coefficient)	0.1	eV	[5]
γ_C	exponent temperature change (Auger recombination coefficient)	0		[45]
T_{ref}	reference temperature for change of gain dispersion peak frequency	310	K	
$\partial_T \omega$	change of gain dispersion peak frequency with temperature	$-0.12 \cdot 10^{12}$	$(sK)^{-1}$	

1 obtained from a fit to results of a microscopic gain model [56]

3 validated by fit to experimental data

Table A.6: Simulation parameters for structure III

symbol	parameter	value	unit	reference
n_g	group refractive index	3.9		
$\bar{n}$	reference refractive index	3.4		
λ	central wavelength	$970 \cdot 10^{-9}$	m	
r_0	rear facet field amplitude reflectivity	$\sqrt{0.95}$		
r_L	front facet field amplitude reflectivity	$\sqrt{0.01}$		
d	thickness of active region	$7 \cdot 10^{-9}$	m	
Γ	confinement factor in active region	$6.6 \cdot 10^{-3}$		
κ	DBR coupling coefficient	0	m^{-1}	
g'	differential modal gain	2300	m^{-1}	1,4
N_{tr}	transparency carrier density	$1.7 \cdot 10^{24}$	m^{-3}	1
N_0	gain clamping carrier density	$4 \cdot 10^{23}$	m^{-3}	1
ϵ_s	gain compression factor	$6.6 \cdot 10^{-26}$	m^3	[49][4]
g_r	amplitude of Lorentzian	4500	m^{-1}	1
γ	Lorentzian half width at half maximum	$1.5 \cdot 10^{13}$	s^{-1}	1
$\delta\omega$	detuning of Lorentzian	0	s^{-1}	1
$\alpha_{0,eff}$	effective internal background absorption	40	m^{-1}	2
f_N	cross section for free carrier absorption	$16 \cdot 10^{-22}$	m^2	2
β'	effective two-photon absorption coefficient	$1.2 \cdot 10^{-4}$	W^{-1}	
n'_N	carrier density induced differential index	$4 \cdot 10^{-32}$	m^3	1
n'_T	temperature induced differential index	$2.5 \cdot 10^{-4}$	K^{-1}	[122]
n'_2	effective Kerr coefficient	$-1.1 \cdot 10^{-11}$	mW^{-1}	
E_g	band gap energy	1.36	eV	
N_c	conduction band density of states	$1 \cdot 10^{24}$	m^{-3}	
N_v	valence band density of states	$5 \cdot 10^{24}$	m^{-3}	
n_0	equilibrium electron carrier density	$1 \cdot 10^{22}$	m^{-3}	
p_0	equilibrium hole carrier density	$4 \cdot 10^{7}$	m^{-3}	
μ_p	hole mobility in active region	$300 \cdot 10^{-4}$	$m^2(Vs)^{-1}$	
A	Shockley Read Hall recomb. coefficient	$9.1 \cdot 10^{8}$	s^{-1}	
B	spontaneous emission recomb. coefficient	$1 \cdot 10^{-16}$	m^3s^{-1}	[60]
C	Auger recomb. coefficient	$2 \cdot 10^{-42}$	m^6s^{-1}	[60]

[1] obtained from a fit to results of a microscopic gain model [56]

[2] free carrier absorption cross sections [120]: $f_n = 4 \cdot 10^{-22}$ m^2 (n-doped layers), $f_p = 12 \cdot 10^{-22}$ m^2 (p-doped layers), $f_n + f_p$ (active region)

[4] includes active region confinement factor

Table A.7: Simulation parameters for electrical model (structure III)

	layer thickness d_i (nm)	electrical conductivity σ_i σ_i ($\Omega^{-1}m^{-1}$)
contact	130	26680
cladding	445	1660
confinement	770	163

Appendix B

Nonlinear susceptibility

The energy gaps of the confinement and cladding layers of a laser structure are larger than the energy of the photons generated by stimulated emission. In this situation, the photon energy is too small to allow absorption of single-photons due to transitions of electrons between the valence and conduction bands. However, transitions involving two photons are still possible which result in a third-order susceptibility.

Assuming an isotropic medium and an instantaneous nonlinear response the nonlinear polarisation density reads [1, 49]

$$P_{\mathrm{NL}}(\vec{r},t) \approx \varepsilon_0 \frac{3}{4} \chi^{(3)}(\vec{r},t) |E(\vec{r},t)|^2 E(\vec{r},t) \tag{B.1}$$

with $E(\vec{r},t)$ from Eq. (2.2) and the third-order susceptibility $\chi^{(3)}$. The commonly used optical Kerr coefficient n_2 and two-photon absorption coefficient β are related to the time-averaged intensity $I = \varepsilon_0 \bar{n} c |E|^2/2$ and defined as

$$\Delta n_2 \equiv \frac{3}{4 n_{\mathrm{r}} \varepsilon_0 \bar{n} c} \chi^{(3)} = n_2 - i \frac{\beta}{2k_0} \tag{B.2}$$

where n_2 has the unit m^2/W and β the unit m/W.

An approximate expression for the dispersion of the two-photon absorption coefficient β for direct transitions has been given in [50],

$$\beta(\omega) = K \frac{\sqrt{E_{\mathrm{p}}}}{n_{\mathrm{r}}^2 E_{\mathrm{g}}^3} F_2 \left(\frac{\hbar\omega}{E_{\mathrm{g}}} \right) \tag{B.3}$$

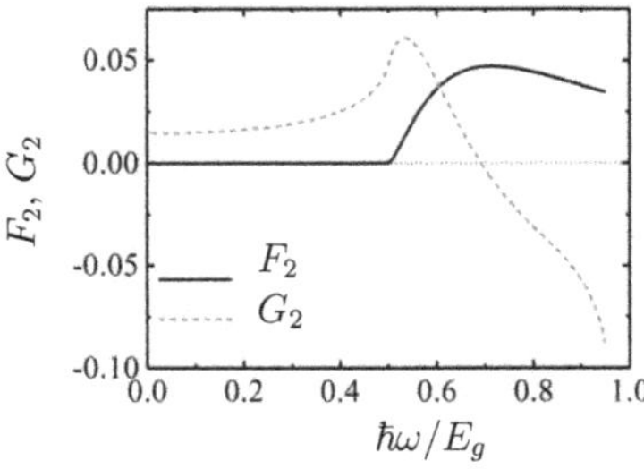

Figure B.1: Functions F_2 and G_2 used in (B.3) and (B.5).

where E_p is the energy equivalent of the momentum matrix element for direct transitions between the valence and conduction bands, E_g is the (direct) energy gap and

$$F_2(x) = \frac{(2x-1)^{3/2}}{(2x)^5} \quad \text{for} \quad 2x > 1. \tag{B.4}$$

The factor K can be considered to be a free parameter. The two-photon absorption coefficient $\beta = 260$ m/TW experimentally determined for GaAs at a wavelength of $\lambda = 1064$ nm [1] is obtained with $K = 41200$ m/TW $\times$ eV$^3/\sqrt{\text{eV}}$ using $E_\mathrm{g} = 1.42$ eV, $E_\mathrm{p} = 26.1$ eV and $n_\mathrm{r} = 3.48$. Note the E_g^{-3} dependence of β.

Two-photon absorption as well as Raman and Stark effects result also in an intensity-dependent contribution to the real part of the refractive index expressed as the optical Kerr coefficient n_2,

$$n_2(\omega) = \tilde{K}\frac{\hbar c\sqrt{E_\mathrm{p}}}{2n_\mathrm{r}^2 E_\mathrm{g}^4} G_2\left(\frac{\hbar\omega}{E_\mathrm{g}}\right) \tag{B.5}$$

where the function $G_2(\hbar\omega/E_\mathrm{g})$ given in [50] is shown in Fig. B.1 together with the function $F_2(\hbar\omega/E_\mathrm{g})$.

The factor $\tilde{K}$ can again be considered as a free parameter. The optical Kerr coefficient $n_2 = -4.1 \cdot 10^{-17}$ m/W experimentally determined for GaAs at a wavelength of $\lambda = 1064$ nm [1] is obtained with $\tilde{K} = 7.212 \cdot 10^{11}$ m/W $\times$ 1/(Ws m) $\times$ eV$^4/\sqrt{\text{eV}}$.

The dependencies of β and n_2 of $Al_xGa_{1-x}As$ on λ are shown in Figs. B.2(a) and B.2(b), respectively. The two-photon absorption increases from $\beta \propto 10$ cm/GW to $\beta \propto 25$ cm/GW if the Al composition is decreased from $x = 0.4$ to $x = 0$ due to the decrease of E_g. It shows a non-monotonous dependence on the wavelength given by $F_2(\lambda)$. Further investigations of two-photon absorption on the output power can be found in Chapter 5. The optical Kerr coefficient decreases with increasing x and changes its sign in dependence on λ at higher Al compositions within the wavelength range investigated. For smaller Al compositions and wavelengths n_2 is negative, but for larger Al compositions and wavelengths n_2 is positive. For $\lambda = 980$ nm and relevant compositions, $|n_2| < 2 \cdot 10^{-4}$ cm^2/GW. Under CW operation the maximum power density of broad-area lasers which are state of the art during the writing of this thesis

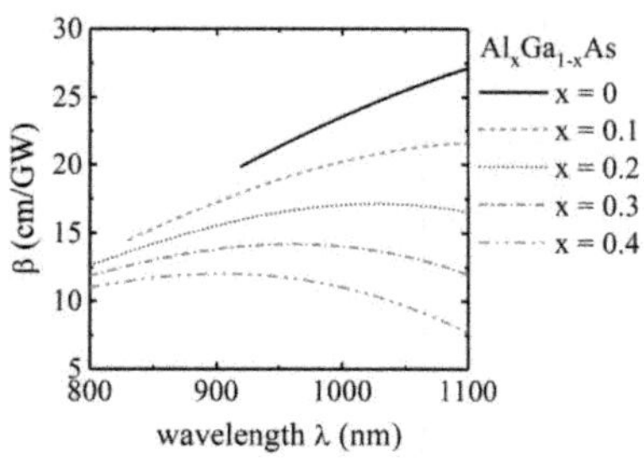

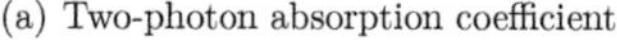
(a) Two-photon absorption coefficient.

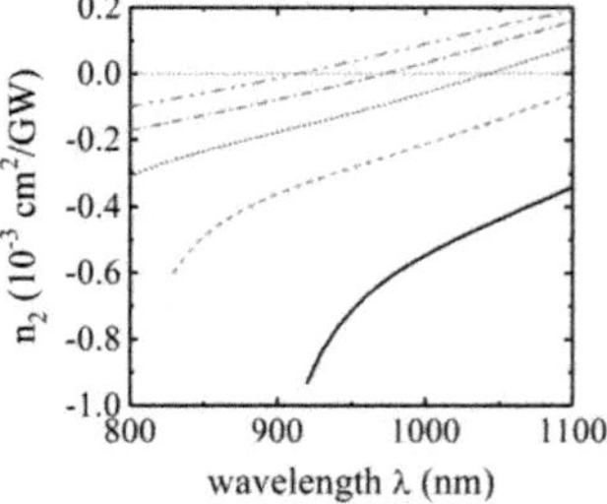

(b) Optical Kerr coefficient.

Figure B.2: Nonlinear coefficients calculated by (B.3) and (B.5) versus wavelength for different Al compositions x in $Al_xGa_{1-x}As$.

is of the order 10^{-2} GW/cm^2. Hence the resulting index change $|\Delta n| < 4 \cdot 10^{-6}$ has a negligible impact on the optical field.

It should be noted that in non-isotropic media the relation between the nonlinear polarisation and the electric field is more complicated than given in (B.1). Hence the two-photon absorption and the optical Kerr effect depend on the crystallographic orientation and on the polarisation direction of the optical field.

Appendix C

The Fermi integral $F_{1/2}$

The Fermi function $f(E)$ gives the occupation probability of a state of energy E at temperature T in thermodynamic equilibrium,

$$f(E) = \frac{1}{1 + \exp(\frac{E-E_F}{k_B T})} \tag{C.1}$$

with the Boltzmann constant k_B and Fermi energy E_F, denoting the energy at which the occupation probability is 1/2. The electron density n in the conduction and hole density p in the valence band can be obtained by integrating the product of density of states with the occupation probability and integrating over all energy states. For parabolic bands the density of states in conduction and valence band is given by

$$\rho_c(E) = \frac{2N_c}{\sqrt{\pi} k_B T}\sqrt{\frac{E - E_c}{k_B T}} \quad \text{and} \quad \rho_v(E) = \frac{2N_v}{\sqrt{\pi} k_B T}\sqrt{\frac{E_v - E}{k_B T}} \tag{C.2}$$

with the effective density of states

$$N_c = 2\left(\frac{2\pi m_n^* k_B T}{h^2}\right)^{3/2} \quad \text{and} \quad N_v = 2\left(\frac{2\pi m_p^* k_B T}{h^2}\right)^{3/2}, \tag{C.3}$$

where m_n^* and m_p^* are the effective masses of electrons and holes respectively. Thus the electron and hole densities can be obtained by

$$n = \int_{E_c}^{\infty} \rho_c(E) f(E) dE \quad \text{and} \quad p = \int_{-\infty}^{E_v} \rho_v(E)(1 - f(E)) dE. \tag{C.4}$$

By setting

$$y = n/N_c, \quad x = (e\varphi_n + e\varphi - E_c)/k_B T \quad \text{and} \quad \xi = (E - E_c)/k_B T \quad \text{for electrons,} \tag{C.5}$$

$$y = n/N_v, \quad x = (E_v + e\varphi_p - e\varphi)/k_B T \quad \text{and} \quad \xi = (E_v - E)/k_B T \quad \text{for holes,} \tag{C.6}$$

respectively, those equations can be written as

$$y = F_{1/2}(x) \tag{C.7}$$

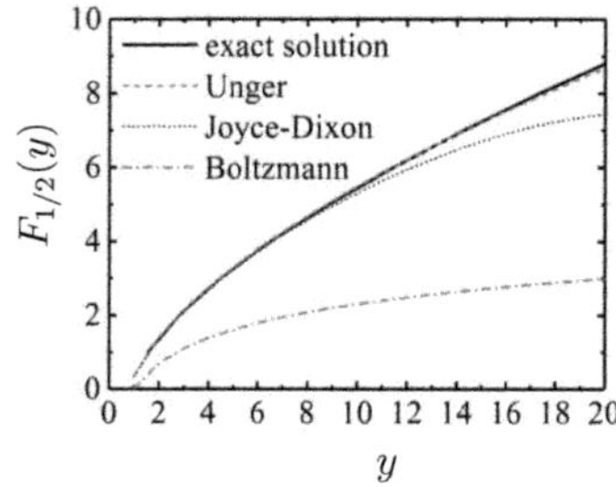

Figure C.1: Solution of the Fermi integral $x = F_{1/2}(y)$ as function of $y = n/N_\mathrm{c}$ or $y = p/N_\mathrm{v}$ and its approximations (Boltzmann, Joyce-Dixon and Unger approximation).

with the Fermi integral of one-half $F_{1/2}$ [123],

$$F_{1/2}(x) = \frac{2}{\sqrt{\pi}} \int_0^\infty \frac{\sqrt{\xi}}{\exp(\xi - x) + 1}\, dy. \tag{C.8}$$

The Fermi integral can't be evaluated analytically, but has to be approximated. In the high temperature regime (Boltzmann limit) the inverse Fermi-integral of one-half $F_{1/2}^{\mathrm{inv}}(y)$ can be approximated by

$$F_{1/2}^{\mathrm{inv}}(y) \approx \ln(y). \tag{C.9}$$

As the Boltzmann limit is only valid at small carrier densities (Fig. C) various more suitable approximations exist for the Fermi-integral [123], such as the Joyce-Dixon approximation [63],

$$F_{1/2}^{\mathrm{inv}}(y) \approx \ln y + a_1 y + a_2 y^2 + a_3 y^3 + a_4 y^4 \tag{C.10}$$

with the coefficients $a_1 = 1/\sqrt{8}$, $a_2 \approx -4.95009 \cdot 10^{-3}$, $a_3 \approx 1.43886 \cdot 10^{-4}$ and $a_4 \approx -4.42563 \cdot 10^{-6}$ or the Unger approximation [124],

$$F_{1/2}^{\mathrm{inv}}(y) \approx \ln \left[\exp\left(\frac{1}{2a} \left\{ \sqrt{1 + 4ay} - 1 \right\} \right) - 1 \right] \tag{C.11}$$

with the adjustment parameter $a = 0.15$. Those equations are good approximations to the Fermi-integral also at higher carrier densities (Fig. C).

Within reasonable ranges for the carrier densities of up to $y < 10$ both the Joyce-Dixon and Unger approximation are well valid. At extremely high carrier densities $y > 10$ the Joyce-Dixon approximation diverges from the exact solution, whereas the Unger approximation still yields acceptable values.

Appendix D

Numerical schemes

The numerical modelling and implementation of the governing equations was performed at the Weierstrass-institute (WIAS) by M. Radziunas and J. Fuhrmann. Here a short summary of the numeric schemes as published in [23, 42, 43, 44] will be given.

In Section D.1 the electro-optic solver is discussed. The dynamic optical traveling-wave equations for the forward- and backward-propagating field $u^{\pm}$ (2.24) are solved in the (x, z)-domain shown in Fig. 1.1(b) by a split-step method using for temporal-longitudinal propagation a finite difference scheme and for the lateral diffraction of the fields a discrete fast Fourier transform algorithm. The dynamic carrier rate equation (3.39) is approximated in the same domain by a finite difference scheme. Due to the complexity of the problem and large simulation domain, the tuning of the model with respect to one or several parameters for device optimization is only possible by means of parallel computing (Section D.1.3).

The inhomogeneous current spreading (3.43) and static heat transport problem (4.45) are solved in several lateral-vertical (x, y)-cross-sections of the device, exemplary shown in Fig. 1.1(c) and (d), respectively, using a finite volume discretization method [125] as discussed in Section D.2. Due to the linearity of the current spreading problem, it can be solved in an initiation step, and then used at each time step of the electro-optic solver, making the efficient implementation possible. In contrast to the current spreading solver, the static heat transport problem is solved only once before resolving time transients with the dynamic electro-optical solver.

D.1 Traveling wave and carrier rate equations

The optical traveling wave equations for the forward- and backward-propagating field $u^{\pm}$ (2.24) and the diffusion equation for the carrier density N (3.39) are solved in the (x, z)-domain shown in Fig. 1.1(b). At the lateral edges of the simulation domain $[-X, X]$ periodic boundary conditions (2.11) and at the device facets $[0, L]$ reflecting boundary conditions (2.10) are employed.

The group velocity defines the fixed ratio of the longitudinal Δz and temporal Δt steps of the mesh, $v_g = c_0/n_g = \Delta z/\Delta t$, such that the mesh diagonals are given by the characteristic lines $z \pm c_0/n_g t = \text{const.}$ [43]. This can be exploited in the scaling of the mesh, so that the factor $c_0/n_g \to 1$ and the longitudinal space and time mesh steps of the scaled problem are equal to $\tau = L/N_z$, which is defined by the number of longitudinal mesh steps N_z. The spatial and temporal grid is uniform and x, z, and t for the optical field is discretized on the mesh $(\omega_x \times \omega_z \times \omega_t)$ where

$$\omega_x = \{x_s = s\Delta x, \quad s = -N_x/2, ..., N_x/2 - 1, \quad \Delta x = 2X/N_x\}, \tag{D.1}$$
$$\omega_z = \{z_k = k\tau, \quad k = 0, ..., N_z, \quad \tau = L/N_z\}, \tag{D.2}$$
$$\omega_t = \{t_j = j\tau, \quad j = 0, ..., t_{\max}/\tau\}. \tag{D.3}$$

The polarization functions are approximated on the related meshes with longitudinal $\tau/2$-shifts $(\omega_x \times \omega_z^* \times \omega_t)$, and the carrier density functions on the related meshes with longitudinal-temporal $\tau/2$-shifts $(\omega_x \times \omega_z^* \times \omega_t^*)$, with

$$\omega_z^* = \{z_{k-1/2} = (k - 1/2)\tau, \quad k = 1, ..., N_z\}, \tag{D.4}$$
$$\omega_t^* = \{t_{j-1/2} = (j - 1/2)\tau, \quad j = 1, ..., t_{\max}/\tau\}. \tag{D.5}$$

For the simulation a fine spatial and temporal grid is needed. For example in the analysis performed in Section 6.4 lateral and longitudinal discretization step sizes of $\Delta x = 0.2$ µm and $\Delta z = \tau = 5$ µm are chosen, respectively. The total lateral and longitudinal widths of the simulation domain $2X = 500$ µm and $L = 4$ mm in this case result in a $N_x \times N_z = 2000 \times 400$ spatial mesh. In this case, the unscaled time step is $\Delta t = \Delta z / v_g = 65$ fs and for 1 ns simulation time approximately $1.5 \cdot 10^4$ time iterations have to be performed. The numerical dimensions of other simulations discussed in this thesis are comparable.

D.1.1 Split step Fourier method for solution of traveling wave equations

To resolve the field equations (2.24) within the time interval $[t_j, t_{j+1}]$ a splitting scheme is utilized to separately solve the nonlinear interaction,

$$\partial_t u'^{\pm} = \mp \partial_z u'^{\pm} - i\Delta\beta^{\pm} u'^{\pm} - \frac{g_r(x,z)n_r}{2\bar{n}}(u'^{\pm} - p^{\pm}) - ik_0\kappa u'^{\mp} + f_{\mathrm{sp}}^{\pm}, \quad u'(t_j) = u(t_j), \tag{D.6}$$

with the propagation function $\Delta\beta^{\pm} = k_0(\Delta n_{\mathrm{eff}} + \Delta n_{2,\mathrm{eff}}^{\pm})$ from the diffraction part of the problem,

$$\partial_t u''^{\pm} = -\frac{i}{2\bar{n}k_0}\partial_x^2 u''^{\pm}, \quad u''(t_j) = u'(t_j + \tau), \quad u(t_j + \tau) = u''(t_j + \tau). \tag{D.7}$$

In this way, there is no need to solve large systems of linear equations by the factorization algorithm.

At the beginning of the j-th time iteration $N_x \times (N_z + 1)$ values for the discrete optical fields u^+ and u^- are defined on the mesh $(\omega_x \times \omega_z \times \omega_t)$, polarization fields p^+ and p^- on the the mesh $(\omega_x \times \omega_z^* \times \omega_t)$ and carrier densities N on the $(\omega_x \times \omega_z^* \times \omega_t^*)$. The discrete operators will be denoted as $A_{s,k}^j = A(x_s, z_k, t_j)$. Furthermore in the following a number of finite difference approximations are used,

$$A_{s,k}^{j+1/2} \approx (A_{s,k}^{j+1} + A_{s,k}^j)/2, \tag{D.8}$$
$$\partial_t^h A_{s,k}^{j+1/2} \approx (A_{s,k}^{j+1} - A_{s,k}^j)/\tau, \tag{D.9}$$
$$u_{s,k-1/2}^{\pm,j+1/2} \approx (u_{s,k-1/2\pm1/2}^{\pm,j+1} - u_{s,k-1/2\mp1/2}^{\pm,j})/2, \tag{D.10}$$
$$\partial_{t\pm z}^h u_{k-1/2}^{\pm,j+1/2} \approx (u_{k-1/2\pm1/2}^{\pm,j+1} - u_{k-1/2\mp1/2}^{\pm,j})/\tau. \tag{D.11}$$

First the diffractionless field and polarization equations (D.6) are approximated on the time layer $j + 1/2$ and solved for the time layer $j + 1$. The propagation function $\Delta\beta^{\pm}$ is approximated on the staggered mesh point $(x_s, z_{k-1/2}, t_{j+1/2})$ from the known carrier density $N_{k-1/2}^{j+1/2}$ and approximated photon density $|u_{k-1/2}^{\pm,j+1/2}|^2 = |(u_{k-1/2\pm1/2}^{\pm,j+1} + u_{k-1/2\pm1/2}^{\pm,j})/2|^2$, as

$$\Delta\beta^{\pm}(N_{k-1/2}^{j+1/2}, |u_{k-1/2}^{\pm,j+1/2}|^2) \approx \Delta\beta^{\pm}(x_s, z_{k-1/2}, t_{j+1/2}). \tag{D.12}$$

The polarization-induced correction functions

$$\Delta^{\pm} = -g_r(u'^{\pm} - p^{\pm}) \tag{D.13}$$

appearing in the field equations are approximated at the same point $(x_s, z_{k-1/2}, t_{j+1/2})$ by the approximated field $u_{k-1/2}^{\prime\pm,j+1/2}$ and initially known discrete polarizations $p_{k-1/2}^{\pm,j+1/2}$ in the Crank-Nicolson approximation,

$$p_{k-1/2}^{\pm,j+1/2} = a(\tau)p_{k-1/2}^{\pm,j} + \frac{\tau\gamma a(\tau)}{2} u_{k-1/2}^{\pm,j+1/2}, \tag{D.14}$$

with $a(\tau) = 2/[2 + \tau(\gamma - i\delta\omega)]$, resulting in

$$\Delta^{\pm}(x_s, z_{k-1/2}, t_{j+1/2}) \approx -g_r(u'^{\pm,j+1/2} - p^{\pm,j+1/2}) \tag{D.15}$$

$$= g_r\Big[\frac{\gamma\tau a(\tau)}{2} - 1\Big]u_{k-1/2}^{\prime\pm,j+1/2} + g_r a(\tau)p_{k-1/2}^{\pm,j}. \tag{D.16}$$

Inserting the above approximations into (D.6) yields

$$\partial_{t\pm z}^{h} u_{k-1/2}^{\prime\pm,j+1/2} = \tag{D.17}$$

$$- i\Delta\beta^{\pm}u_{k-1/2}^{\prime\pm,j+1/2} - i\kappa u_{k-1/2}^{\prime\mp,j+1/2} + f_{\rm sp}^{\pm} + g_r\Big[\frac{\gamma\tau a(\tau)}{2} - 1\Big]u_{k-1/2}^{\prime\pm,j+1/2} + g_r a(\tau)p_{k-1/2}^{j},$$

where $k = 1, \ldots, N_z$. Inserting (D.10) and (D.11) gives,

$$\Big[i\Delta\beta^{\pm} + \frac{2}{\tau} - g_r\Big(\frac{\gamma\tau a(\tau)}{2} - 1\Big)\Big]u_{k-1/2}^{\prime\pm,j+1/2} + i\kappa u_{k-1/2}^{\prime\mp,j+1/2} = \tag{D.18}$$

$$+ \frac{2}{\tau}u_{k-1/2\mp1/2}^{\pm,j+1/2} + g_r a(\tau)p_{k-1/2}^{j} + f_{\rm sp}^{\pm},$$

which can be easily pair-wise resolved with respect to $u'^{+,j+1/2}$ and $u'^{-,j+1/2}$ defined for ω_z^* and $t_{j+1/2}$. The field values $u_k^{\prime j+1}$ at the new t_{j+1}-time layer are given by (D.10) as

$$u_{s,k-1/2\pm1/2}^{\pm\prime j+1} = 2u_{s,k-1/2}^{\prime\pm,j+1/2} - u_{s,k-1/2\mp1/2}^{\pm,j} \quad \text{for} \quad k = 1, \ldots, N_z, \tag{D.19}$$

Before switching to the next splitting step that accounts for the field diffraction, first the polarization functions on the new time layer t_{j+1} are reconstructed from $u_{k-1/2}^{\prime\pm,j+1/2}$ and $p_{k-1/2}^{\pm,j}$ using finite differences,

$$\frac{p_{k-1/2}^{\pm,j+1} - p_{k-1/2}^{\pm,j}}{\tau} = \gamma u_{k-1/2}^{\prime\pm,j+1/2} - (\gamma - i\delta\omega)\frac{p_{k-1/2}^{\pm,j+1} - p_{k-1/2}^{\pm,j}}{2} \tag{D.20}$$

$$\Rightarrow p_{k-1/2}^{\pm,j+1} = (2a(\tau) - 1)p_{k-1/2}^{\pm,j} + 2\frac{\gamma\tau a(\tau)}{2}u_{k-1/2}^{\prime\pm,j+1/2} \tag{D.21}$$

and the functions $\mathrm{Re}\big[u_{k-1/2}^{\prime\pm*,j+1/2}\cdot\Delta_{k-1/2}^{j+1/2}\big]$, with $u_{k-1/2}^{\prime\pm,j+1/2}$ from (D.18) and $\Delta_{k-1/2}^{\pm,j+1/2}$ from (D.16), are collected that are needed to be used later when resolving the carrier rate equation (D.27). Alternatively, the polarization functions can be also solved by an exponentially weighted scheme as described in [23, 126].

In the second stage field diffraction is taken into account by solving Eq. (D.7) with the Fourier method. In the approach it is assumed that $u_s^{\prime\prime\pm}(z,t) := u^{\prime\prime\pm}(z,x_s,t)$ can be expressed as a linear combination of the orthonormal grid-functions $e^{i\pi l x_s/X}, l = -N_x/2, ..., N_x/2-1$,

$$u_s^{\prime\prime\pm}(z,t) = \Big[\mathscr{F}^{-1}\Big(\hat{u}_l^{\pm}(z,t)\big|_{l=-N_x/2}^{N_x/2-1}\Big)\Big]_s := \frac{1}{N_x}\sum_{l=-N_x/2}^{N_x/2-1}\hat{u}_l^{\pm}(z,t)e^{i\pi l x_s/X}, \tag{D.22}$$

where the Fourier coefficients are defines as

$$\hat{u}_l^{\pm}(z,t) = \Big[\mathscr{F}\Big(u_s^{\pm}(z,t)\big|_{s=-N_x/2}^{N_x/2-1}\Big)\Big]_l := \sum_{s=-N_x/2}^{N_x/2-1}u_s^{\prime\prime\pm}(z,t)e^{-i\pi l x_s/X}. \tag{D.23}$$

In this way the approximation of the lateral second order derivative can be written as

$$\partial_x^2 u_s^{\prime\prime\pm}(z,t) \approx \frac{1}{N_x}\sum_{l=-N_x/2}^{N_x/2-1}\Big(-\frac{\pi^2 l^2}{X^2}\Big)\hat{u}_l^{\pm}(z,t)e^{i\pi l x_s/X}. \tag{D.24}$$

To integrate (D.7) the preceding approximations are used and the resulting systems of differential equations in the lateral Fourier domain are solved to gain the expression,

$$u_{m,k}^{\prime\prime\pm,j+1} = \frac{1}{N_x}\sum_{\ell=-N_x/2}^{N_x/2-1}\left[e^{i\frac{D_u\pi^2\ell^2}{X^2}\tau}\sum_{s=-N_x/2}^{N_x/2-1}u_{s,k}^{\prime\prime\pm}e^{-i\frac{2\pi\ell s}{N_x}}\right]e^{i\frac{2\pi\ell m}{N_x}},\quad u_{s,k}^{\prime\prime\pm} = u_{s,k}^{\prime\pm,j+1}, \tag{D.25}$$

$$m = -N_x/2,\dots,N_x/2-1,\quad u_{m,k}^{\pm,j+1} = u_{m,k}^{\prime\prime\pm,j+1},\quad k = \frac{1}{2}\pm\frac{1}{2},\dots,N_z-\frac{1}{2}\pm\frac{1}{2},$$

for the field diffraction, which completes the numerical description of the traveling-wave equation.

In the final step the missing fields at $k = 0$ and $k = N_z$ are defined using the reflecting boundary conditions (2.10), i.e.

$$u_{s,0}^{+,j+1} = r_0 u_{s,0}^{-,j+1} \quad\text{and}\quad u_{s,N_z}^{-,j+1} = r_L e^{-i2\bar{n}k_0 L}u_{s,N_z}^{+,j+1}. \tag{D.26}$$

At the end of these iterations the field functions $u^{\pm,j+1}$ at the $j+1$ time layer are fully defined, so that the photon densities $|u_{k-1/2}^{\pm,j+1}|^2$ can be calculated at this time layer, which are needed for the approximation of the carrier rate equation (D.27).

D.1.2 Finite difference scheme for solution of carrier rate equations

We will now turn to the solution of the carrier rate equations. In a previous version [23] the carrier rate equations were solved by the split step Fourier method, however as it relies on a laterally-constant carrier diffusion coefficients [43] this approach is replaced by the Crank- Nicolson-type [127] finite difference scheme. In contrast to the Fourier based method, the finite difference approach not only allows a laterally varying carrier diffusion but also different lateral boundary conditions, such as periodic and Neumann boundary conditions.

For the finite difference scheme the carrier rate equation,

$$\partial_t N = \partial_x(D_{\text{eff}}\partial_x N) + \frac{j}{ed} - R(N) - g_{\text{eff}}\|u\|^2 - \sum_{\nu=+,-} \text{Re}\left(u^{\nu*} \cdot \Delta^{\nu}\right), \tag{D.27}$$

has to be solved for the grid point $(x_s, z_{k-1/2}, t_{j+1})$. To do so first the injection current density $j^{j+1/2}_{s,k-1/2}$ (for the solution of the inhomogeneous current spreading model see section (D.2)) and diffusion coefficient $D^{j+1/2}_{\text{eff},s,k-1/2}$ are calculated from the known values of the carrier density $N^{j+1/2}_{s,k-1/2}$. For each longitudinal position $z_{k-1/2}$ the effective gain and recombination rate are approximated for the $j+1$ time layer by

$$g^{j+1}_{\text{eff}} := g_{\text{eff}}(N^{j+1/2}, \|u^{j+1}\|^2) + \partial_N g_{\text{eff}}(N^{j+1/2}, \|u^{j+1}\|^2)\left(N^{j+1} - N^{j+1/2}\right), \tag{D.28}$$

$$R^{j+1} := R(N^{j+1/2}, \|u^{j+1}\|^2) + \partial_N R(N^{j+1/2}, \|u^{j+1}\|^2)\left(N^{j+1} - N^{j+1/2}\right). \tag{D.29}$$

Furthermore the following approximations are needed,

$$N^{j+1}_s \approx (N^{j+3/2}_s + N^{j+1/2}_s)/2, \tag{D.30}$$

$$\partial^h_t N^{j+1}_s \approx (N^{j+3/2}_s - N^{j+1/2}_s)/\tau, \tag{D.31}$$

$$\left[\partial_x(D_{\text{eff}}\partial_x N)\right]_s \approx \frac{(D_{\text{eff},s+1} + D_{\text{eff},s})(N_{s+1} - N_s) - (D_{\text{eff},s} + D_{\text{eff},s-1})(N_s - N_{s-1})}{2(\Delta x)^2}. \tag{D.32}$$

Resolving the carrier rate equation for the mesh point $z_{k-1/2}$ gives,

$$\begin{aligned}\partial^h_t N^{j+1}_s = \left[\partial_x\left(D^{j+1/2}_{\text{eff}}\partial_x N^{j+1}\right)\right]_s + \frac{j^{j+1/2}_s}{ed} - R^{j+1}_s - g^{j+1}_{\text{eff},s}\|u^{j+1}_s\|^2 \\ - \sum_{\nu=+,-} \text{Re}\left(u'^{\nu*,j+1/2}_s \cdot \Delta^{\nu,j+1/2}_s\right),\end{aligned} \tag{D.33}$$

and substituting (D.28)-(D.32) result in,

$$\begin{aligned}\Rightarrow -\frac{D^{j+1/2}_{\text{eff},s} + D^{j+1/2}_{\text{eff},s-1}}{2(\Delta x)^2} N^{j+1}_{s-1} + \left(\mathcal{D}_s + \frac{D^{j+1/2}_{\text{eff},s+1} + 2D^{j+1/2}_{\text{eff},s} + D^{j+1/2}_{\text{eff},s-1}}{2(\Delta x)^2}\right) N^{j+1}_s \\ - \frac{(D^{j+1/2}_{\text{eff},s+1} + D^{j+1/2}_{\text{eff},s})}{2(\Delta x)^2} N^{j+1}_{s+1} = \mathcal{F}_s,\end{aligned} \tag{D.34}$$

$$\begin{aligned}\text{where}\quad \mathcal{D}_s =& \frac{2}{\tau} + \partial_N R(N^{j+1/2}_s) + \partial_N g_{\text{eff}}(N^{j+1/2}_s, \|u^{j+1}_s\|^2)\|u^{j+1}_s\|^2, \\ \mathcal{F}_s =& \mathcal{D}_s N^{j+1/2}_s + \frac{j^{j+1/2}_s}{ed} - R(N^{j+1/2}_s) - g_{\text{eff}}(N^{j+1/2}_s, \|u^{j+1}_s\|^2)\|u^{j+1}_s\|^2 \\ &- \sum_{\nu=+,-} \text{Re}\left(u'^{\nu*,j+1/2}_s \cdot \Delta^{\nu,j+1/2}_s\right).\end{aligned}$$

The resulting schemes with respect to N^{j+1} are defined by a three-diagonal operator and resolved by a standard fast factorization algorithm.

Note, that for the polarization induced correction functions $\Delta^{\pm,j+1/2}_s$ the same approximations (D.16) are exploited as in the field equations (D.17). This ensures that any induced artificial contributions in the field and carrier rate equations, that might arise from the polarization induced correction, cancel each other.

D.1.3 Parallelization

For device optimization of BA laser devices the fine spatial and temporal grid has to be merged with an efficient algorithm. For this, the parallelization of the algorithms is indispensable and done by domain decomposition using the distributed memory paradigm.

As the Fourier method is employed to solve the diffraction contribution in the traveling-wave equation, it is more difficult to split the domain in lateral direction. Accordingly the full problem defined by N_x and N_z is split along the longitudinal z-coordinate into different subdomains corresponding to the chosen number of processes q. The domains are split along the mesh points z_{k_ι} with $\iota = 1, ..., q$ and the model equations are solved on the longitudinal subdomain $[z_{k_{\iota-1}} z_{k_\iota}]$. The advantage of the staggered mesh choice for the approximation of the polarization and carrier density functions is, that before the execution of each time iteration step, any two adjacent subdomain processes exchange only the necessary information for the complex forward- and backward-traveling fields $u^{\pm}_{k_\iota}$. The discrete polarization and carrier density functions, however, are attributed to the unique process. The exchange of information leads to a slight slow down for a higher number of processes so that the speed up using 30 processes is slightly more than a factor of 20 [43].

Accordingly by parallelizing the algorithms for the above example with a spatial mesh of $N_x \times N_z = 2000 \times 800$ with lateral and longitudinal discretization steps of $\Delta x = 0.2$ µm and $\Delta z = 5$ µm, which would require 3.5 h for the simulation of a 1 ns long transient using one single process, the simulation can be performed in less than 10 minutes on the multicore server at the Weierstrass Institute in Berlin using 30 processes. In [23, 43] more details on the parallelizing algorithms and a more detailed performance analysis on the parallel computer cluster at the Weierstrass Institute in Berlin is given.

D.2 Current spreading and heat transport solvers

Both the inhomogeneous current spreading (3.43) and static heat transport problem (4.45) are solved in N_z lateral-vertical $(x, y) := (x, y, z_0)$ slabs of the simulation domain by means of a finite volume discretization method [125].

The simulation domain is discretized with a suitable triangulation, where the straight interfaces between materials, see Fig. 1.1, are aligned with the edges of the triangulation. Thus, it is automatically ensured that the quasi Fermi potential, injection current density, temperature and heat flux are continuous at all internal layer interfaces Γ_i (for the inhomogeneous current speading problem they are marked green in Fig. 3.3), i.e. $\varphi_p|_{(x,y)\in\Gamma_i^+} = \varphi_p|_{(x,y)\in\Gamma_i^-}$, $\sigma\partial_n\varphi|_{(x,y)\in\Gamma_i^+} = \sigma\partial_n\varphi|_{(x,y)\in\Gamma_i^-}$, $T|_{(x,y)\in\Gamma_i^+} = T|_{(x,y)\in\Gamma_i^-}$, and $\kappa\partial_n T|_{(x,y)\in\Gamma_i^+} = \kappa\partial_n T|_{(x,y)\in\Gamma_i^-}$, respectively, where $\Gamma_i^{\pm}$ denote different sides of the internal layer interfaces [44]. The corner points of the triangulation are the central vectors $\vec{r}_i$ of Voronoi cells with control volume V_i.

D.2.1 Current spreading solver

The inhomogeneous current spreading problem (3.43) parametrized by the longitudinal position z_0,

$$\nabla_{(x,y)} \cdot \Big(\sigma(x, y)\nabla_{(x,y)}\varphi_p(x, y)\Big) = 0, \tag{D.35}$$

has to be solved in the domain marked grey in Fig. 3.3 with the piecewise constant conductivity $\sigma(x, y)$. At the metal contact (marked blue in Fig. 3.3) the hole quasi Fermi potential is given by the contact voltage $\varphi_p = U$. At the boundary of p-layer to the active region it is given by the Fermi potential $\varphi_p = \varphi_F(N)$ calculated in the Joyce-Dixon approximation [63] (*cf.* Appendix C). At all other outer device boundaries homogeneous Neumann conditions are imposed, $\partial_n \varphi_p = 0$. The injection current density entering Eq. (3.39) is given by

$$j(x) = \sigma_p(x, y_{\mathrm{ar}}) \partial_y \varphi_p(x, y_{\mathrm{ar}}) \tag{D.36}$$

where y_{ar} gives the y-position of the boundary of active region to the p-layers.

For the solution of (D.35) a finite volume based numerical scheme [125] is used. Eq. (D.35) can be expressed as a surface integral, which is split into contributions from the facets common with the neighboring control volumes,

$$\sum_j \int_{S_{ij}} \left(\sigma \nabla \varphi_p\right) \cdot \vec{n} dS = 0, \tag{D.37}$$

where S_{ij} is the interface between Volume V_i and volume V_j. Using numerical quadrature (D.37) can be written as

$$\sum_j \frac{|S_{ij}|}{|\vec{r}_j - \vec{r}_i|} g(\sigma_{p,i}, \sigma_{p,j}, u_{ij}) = 0, \tag{D.38}$$

where $g(\sigma_{p,i}, \sigma_{p,j}, u_{ij})$ is the flux function, that approximates the scaled normal flux $\sigma \nabla \varphi_p (\vec{r}_j - \vec{r}_i)$ through the common surface S_{ij}, $\sigma_{p,i}$ and $\sigma_{p,j}$ are the values of σ_p in V_i and V_j, respectively, and

$$u_{ij} = \frac{1}{S_{ij}} \int_{S_{ij}} \nabla \varphi_p \cdot \vec{n} dS \tag{D.39}$$

is the average value of the normal flux of $\nabla \varphi_p$ through S_{ij}. The flux function $g(\sigma_{p,i}, \sigma_{p,j}, u_{ij})$ can be expressed in a 1D finite difference formulation, so that the finite volume space discretization results in a linear system of equations,

$$A_h \varphi_{p,h} = 0, \tag{D.40}$$

where A_h and $\varphi_{p,h}$ denote the discrete approximations of the differential operator and the quasi-Fermi potential, respectively, which can be solved by a sparse matrix solver.

Due to its linearity, the current spreading problem (D.35) is solved with a Green function like approach. Using this method a matrix of $N_x \times (N_x + 1)$ dimensions needs to be constructed only once in the initialization step, where N_x is the number of lateral mesh points, *cf.* Eq. (D.1). The matrix contains $N_x + 1$ elementary discrete injected current density vectors j^i, where $i = 0, ..., N_x$, that correspond to the solution of (D.35) with the boundary conditions $(U, \varphi_F(y = y_{\mathrm{ar}})) = (1, 0)$ in the case of $i = 0$, and $(U, \varphi_F(y = y_{\mathrm{ar}})) = (0, \delta_{|x=x_i|}(x))$ for $i = 1, ..., N_x$. Here, $\delta_{|x=x_i|}(x)$ denotes the delta-distribution, that has the value 1 at $x = x_i$ and 0 otherwise.

Then the precalculated Green functions can be used in all consequent simulations to reconstruct the required current distribution for any voltage U and Fermi potential $\varphi_F(y = y_{\mathrm{ar}})$. In order to reduce the number of operations required for the matrix-vector multiplication, the majority of small off-diagonal elements of the $N_x \times (N_x + 1)$ dimensional matrix is ignored. For more information on this approach see [42].

As a result of the large matrices, the multiple matrix-vector multiplications become the main reason of the simulation slow-down when the inhomogeneous current spreading model is included. For the above example with a spatial mesh of $N_x \times N_z = 2000 \times 800$ with lateral and longitudinal discretization steps of $\Delta x = 0.2$ µm and $\Delta z = 5$ µm, applying the inhomogeneous current spreading model in each time iteration causes about 50% slow-down of the calculations. For further information on a more efficient implementation of the current spreading problem which can be applied for a specific lateral device geometry (e.g. the one displayed in Fig. 3.3) and exploits a semi-discrete separation of variables based method see [42].

D.2.2 Heat transport solver

The static heat transport problem parametrized by $z = z_0$,

$$\nabla_{(x,y)} \cdot \left(\kappa_L(x,y)\nabla_{(x,y)}T\right) = -h(x,y), \tag{D.41}$$

is solved in the domain displayed in Fig. 1.1(d) with the piecewise constant thermal conductivity $\kappa_L(x,y)$, subject to the boundaries conditions (4.2). The heat source density is approximated by mean values over the control volumes, i.e. $h_i = \int_{V_i} h(x,y,z_0)dV/|V_i|$.

Thus, using the finite volume discretization, the heat transport equation can be expressed similar to the inhomogeneous current spreading problem by a linear system of equations

$$A_h T_h = -s_h, \tag{D.42}$$

where A_h, s_h and T_h represent the discrete approximations of the differential operator, the source term and the temperature, respectively, which is solved via the LU factorization $A_h = L_h U_h$ into easily invertible upper and lower triangular matrices $T_h = -U_h^{-1} L_h^{-1} s_h$ [44].

The laser structures investigated in this thesis have only few different cross-sections corresponding to different sets of z_0. Thus and because the differential operator is linear the same LU factorization can be used for different z_0. For the investigated laser structures the time required by the thermal solver is consequently performed in the range of $\approx 1-2$ min, which is fast compared to the at least 30 min needed for a standard 5 ns long transient simulation of the electro-optic solver [44].

The vertical size of the control volumes do not exceed a few nanometers within the thin active region, whereas outside of this domain they are much larger. For a representative simulation discussed in [44] the total number of control volumes was $\approx 2 \cdot 10^5$.

Bibliography

[1] R. Boyd, *Nonlinear Optics*. Academic Press, Elsevier, 3rd ed., 2008.

[2] G. H. B. Thompson, "A theory for filamentation in semiconductor lasers including the dependence of dielectric constant on injected carrier density," *Opt. Quantum Electron.*, vol. 4, no. 3, pp. 257–310, 1972.

[3] A. Zeghuzi, M. Radziunas, H.-J. Wünsche, J.-P. Koester, H. Wenzel, U. Bandelow, and A. Knigge, "Traveling wave analysis of non-thermal far-field blooming in high-power broad-area lasers," *IEEE J. Quantum Electron.*, vol. 55, no. 2, p. 2000207, 2019.

[4] A. Zeghuzi, H.-J. Wünsche, H. Wenzel, M. Radziunas, J. Fuhrmann, A. Klehr, U. Bandelow, and A. Knigge, "Time-dependent simulation of thermal lensing in high-power broad-area semiconductor lasers," *IEEE J. Sel. Top. Quantum Electron.*, vol. 25, no. 6, p. 1502310, 2019.

[5] H. Wenzel, P. Crump, A. Pietrzak, X. Wang, G. Erbert, and G. Tränkle, "Theoretical and experimental investigations of the limits to the maximum output power of laser diodes," *New J. Phys.*, vol. 12, no. 8, p. 085007, 2010.

[6] B. S. Ryvkin and E. A. Avrutin, "Effect of carrier loss through waveguide layer recombination on the internal quantum efficiency in large-optical-cavity laser diodes," *J. Appl. Phys.*, vol. 97, p. 113106, 2005.

[7] H. Wenzel, P. Crump, A. Pietrzak, C. Roder, X. Wang, and G. Erbert, "The analysis of factors limiting the maximum output power of broad-area laser diodes," *Opt. Quantum Electron.*, vol. 41, no. 9, pp. 645–652, 2009.

[8] R. S. Tucker and D. J. Pope, "Circuit modeling of the effect of diffusion on damping in a narrow-stripe semiconductor laser," *IEEE J. Quantum Electron.*, vol. 19, no. 7, pp. 1179–1183, 1983.

[9] S. F. Yu, R. G. S. Plumb, L. M. Zhang, M. C. Nowell, and J. E. Carroll, "Large-signal dynamic behavior of distributed-feedback lasers including lateral effects," *IEEE J. Quantum Electron.*, vol. 30, no. 8, pp. 1740–1750, 1994.

[10] M. Dogan, C. P. Michael, Y. Zheng, L. Zhu, and J. H. Jacob, "Two photon absorption in high power broad area laser diodes," *Proc. SPIE*, vol. 8965, p. 89650P, 2014.

[11] E. A. Avrutin and B. S. Ryvkin, "Theory of direct and indirect effect of two-photon absorption on nonlinear optical losses in high power semiconductor lasers," *Semicond. Sci. Technol.*, vol. 32, p. 015004, 2016.

[12] C. Tsai, C. Tsai, R. M. Spencer, Y. Lo, and L. F. Eastman, "Nonlinear gain coefficients in semiconductor lasers: effects of carrier heating," *IEEE J. Quantum Electron.*, vol. 32, no. 2, pp. 201–212, 1996.

[13] J. Wang and H. C. Schweizer, "A quantitative comparison of the classical rate-equation model with the carrier heating model on dynamics of the quantum-well laser: the role of carrier energy relaxation, electron-hole interaction, and Auger effect," *IEEE J. Quantum Electron.*, vol. 33, no. 8, pp. 1350–1359, 1997.

[14] S. O. Slipchenko, Z. N. Sokolova, N. A. Pikhtin, K. S. Borschev, D. A. Vinokurov, and I. S. Tarasov, "Finite time of carrier energy relaxation as a cause of optical-power limitation in semiconductor lasers," *Semiconductors*, vol. 40, no. 8, pp. 990–995, 2006.

[15] A. Zeghuzi, M. Radziunas, H.-J. Wünsche, A. Klehr, H. Wenzel, and A. Knigge, "Influence of nonlinear effects on the characteristics of pulsed high-power broad-area distributed Bragg reflector lasers," *Opt. Quantum Electron.*, vol. 50, no. 88, pp. 1–12, 2018.

[16] P. Crump, S. Böldicke, C. M. Schultz, H. Ekhteraei, H. Wenzel, and G. Erbert, "Experimental and theoretical analysis of the dominant lateral waveguiding mechanism in 975 nm high power broad area diode lasers," *Semicond. Sci. Technol.*, vol. 27, p. 045001, 2012.

[17] S. Rauch, H. Wenzel, M. Radziunas, M. Haas, G. Tränkle, and H. Zimer, "Impact of longitudinal refractive index change on the near-field width of high-power broad-area diode lasers," *Appl. Phys. Lett.*, vol. 110, no. 26, p. 263504, 2017.

[18] G. R. Hadley, J. P. Hohimer, and A. Owyoung, "Comprehensive modeling of diode arrays and broad-area devices with applications to lateral index tailoring," *IEEE J. Quantum Electron.*, vol. 24, no. 11, pp. 2138–2152, 1988.

[19] L. Borruel, S. Sujecki, P. Moreno, J. Wykes, M. Krakowski, B. Sumpf, P. Sewell, S.-C. Auzanneau, H. Wenzel, D. Rodríguez, T. M. Benson, E. C. Larkins, and I. Esquivias, "Quasi-3-D simulation of high-brightness tapered lasers," *IEEE J. Quantum Electron.*, vol. 40, no. 5, pp. 463–472, 2004.

[20] J. Piprek, "Self-consistent analysis of thermal far-field blooming of broad-area laser diodes," *Opt. Quantum Electron.*, vol. 45, no. 7, pp. 581–588, 2013.

[21] J. Piprek and Z. M. S. Li, "On the importance of non-thermal far-field blooming in broad-area high-power laser diodes," *Appl. Phys. Lett.*, vol. 102, no. 221110, pp. 1–4, 2013.

[22] M. Winterfeldt, P. Crump, S. Knigge, A. Maaßdorf, U. Zeimer, and G. Erbert, "High beam quality in broad area lasers via suppression of lateral carrier accumulation," *IEEE Photonics Technol. Lett.*, vol. 27, no. 17, pp. 1809–1812, 2015.

[23] M. Radziunas and R. Ciegis, "Effective numerical algorithm for simulations of beam stabilization in broad area semiconductor lasers and amplifiers," *Math. Model. Anal.*, vol. 19, no. 5, pp. 627–646, 2014.

[24] M. Radziunas, "BALaser: A software tool for simulation of dynamics in broad area semiconductor lasers.." *http://www.wias-berlin.de/software/balaser/*, accessed: 2020-01-01.

[25] M. Niederhoff, W. Heinrich, and P. Russer, "Three-dimensional modelling of high-power laser diodes based on the finite integration beam propagation method," *IEEE MTT-S Int. Microw. Symp. Dig.*, vol. 3, pp. 5–8, 1996.

[26] A. Zeghuzi, H. Wenzel, H.-J. Wünsche, M. Radziunas, U. Bandelow, and A. Knigge, "Modeling of current spreading in high-power broad-area lasers and its impact on the lateral far field divergence," *Proc. SPIE*, vol. 10526, p. 105261H, 2018.

[27] J. R. Marciante and G. P. Agrawal, "Nonlinear mechanisms of filamentation in broad-area semiconductor lasers," *IEEE J. Quantum Electron.*, vol. 32, no. 4, pp. 590–596, 1996.

[28] C. Z. Ning, R. A. Indik, and J. V. Moloney, "Effective bloch equations for semiconductor lasers and amplifiers," *IEEE J. Quantum Electron.*, vol. 33, no. 9, pp. 1543–1550, 1997.

[29] M. Lichtner, M. Radziunas, U. Bandelow, M. Spreemann, and H. Wenzel, "Dynamic simulation of high brightness semiconductor lasers," *Proc. NUSOD*, pp. 65–66, 2008.

[30] A. Pérez-Serrano, J. Javaloyes, and S. Balle, "Spectral delay algebraic equation approach to broad area laser diodes," *IEEE J. Sel. Top. Quantum Electron.*, vol. 19, no. 5, p. 1502808, 2013.

[31] H.-C. Eckstein and U. D. Zeitner, "Modeling electro-optical characteristics of broad area semiconductor lasers based on a quasi-stationary multimode analysis," *Opt. Express*, vol. 21, no. 20, pp. 23231–23240, 2013.

[32] C. Holly, S. Hengesbach, M. Traub, and D. Hoffmann, "Simulation of spectral stabilization of high-power broad-area edge emitting semiconductor lasers," *Opt. Express*, vol. 21, no. 13, pp. 15553–15567, 2013.

[33] P. G. Eliseev, A. G. Glebov, and M. Osiński, "Current self-distribution effect in diode lasers: Analytic criterion and numerical study," *IEEE J. Sel. Top. Quantum Electron.*, vol. 3, no. 2, pp. 499–506, 1997.

[34] E. Gehrig and O. Hess, "Nonequilibrium spatiotemporal dynamics of the Wigner distributions in broad-area semiconductor lasers," *Phys. Rev. A*, vol. 57, no. 3, pp. 2150–2162, 1998.

[35] K. Böhringer and O. Hess, "A full-time-domain approach to spatio-temporal dynamics of semiconductor lasers," *Prog. Quantum Electron.*, vol. 32, pp. 159–246, 2008.

[36] M. Spreemann, M. Lichtner, M. Radziunas, U. Bandelow, and H. Wenzel, "Measurement and simulation of distributed-feedback tapered master-oscillator power amplifiers," *IEEE J. Quantum Electron.*, vol. 45, no. 6, pp. 609–616, 2009.

[37] C. Fiebig, V. Z. Tronciu, M. Lichtner, K. Paschke, and H. Wenzel, "Experimental and numerical study of distributed-Bragg-reflector tapered lasers," *Appl. Phys. B*, vol. 99, no. 1, pp. 209–214, 2010.

[38] M. Spreemann, H. Wenzel, B. Eppich, M. Lichtner, and G. Erbert, "A novel approach to finite-aperture tapered unstable resonator lasers.," *IEEE J. Quantum Electron.*, vol. 47, no. 1, pp. 117–125, 2011.

[39] V. Z. Tronciu, S. Schwertfeger, M. Radziunas, A. Klehr, U. Bandelow, and H. Wenzel, "Numerical simulation of the amplification of picosecond laser pulses in tapered semiconductor amplifiers and comparison with experimental results," *Opt. Commun.*, vol. 285, no. 12, pp. 2897–2904, 2012.

[40] M. Lichtner, V. Z. Tronciu, and A. G. Vladimirov, "Theoretical investigation of striped and non-striped broad area lasers with off-axis feedback," *IEEE J. Quantum Electron.*, vol. 48, no. 3, pp. 353–360, 2012.

[41] M. Radziunas, R. Herrero, M. Botey, and K. Staliunas, "Far-field narrowing in spatially modulated broad-area edge-emitting semiconductor amplifiers," *J. Opt. Soc. Am. B*, vol. 32, no. 5, p. 993, 2015.

[42] M. Radziunas, A. Zeghuzi, J. Fuhrmann, T. Koprucki, H.-J. Wünsche, H. Wenzel, and U. Bandelow, "Efficient coupling of the inhomogeneous current spreading model to the dynamic electro-optical solver for broad-area edge-emitting semiconductor devices," *Opt. Quantum Electron.*, vol. 49, no. 10, pp. 1–8, 2017.

[43] M. Radziunas, "Modeling and simulations of broad-area edge-emitting semiconductor devices," *Int. J. High Perform. Comput. Appl.*, vol. 32, no. 4, pp. 512–522, 2018.

[44] M. Radziunas, J. Fuhrmann, A. Zeghuzi, H.-J. Wünsche, T. Koprucki, C. Brée, H. Wenzel, and U. Bandelow, "Efficient coupling of dynamic electro-optical and heat-transport models for high-power broad-area semiconductor lasers," *Opt. Quantum Electron.*, vol. 51, no. 69, pp. 1–10, 2019.

[45] U. Bandelow, H. Gajewski, and R. Hünlich, "Fabry-Pérot laser: Thermodynamics - based modeling of edge - emitting quantum well lasers," in *Optoelectron. Devices - Adv. Simul. Anal.* (J. Piprek, ed.), ch. 3, pp. 63–85, New York: Springer Science & Business Media, 2005.

[46] G. K. Wachutka, "Rigorous thermodynamic treatment of heat generation and conduction in semiconductor device modeling," *IEEE Trans. Comput. Des.*, vol. 9, no. 11, pp. 1141–1149, 1990.

[47] V. I. Bespalov and V. I. Talanov, "Filamentary structure of light beams in nonlinear liquids," *JETP Lett.*, vol. 3, no. 11, pp. 307–310, 1966.

[48] C. H. Henry and R. F. Kazarinov, "Origins of quantum noise in photonics," *Rev. Mod. Phys.*, vol. 68, no. 3, p. 806, 1996.

[49] H. Wenzel and A. Zeghuzi, "High-Power Lasers," in *Handb. Optoelectron. Device Model. Simul.* (J. Piprek., ed.), ch. 33, pp. 15–58, CRC Press, Taylor & Francis Group, 1rst ed., 2017.

[50] M. Sheik-Bahae, D. C. Hutchings, D. J. Hagan, and E. W. Van Stryland, "Dispersion of bound electronic nonlinear refraction in solids," *IEEE J. Quantum Electron.*, vol. 27, no. 6, pp. 1296–1309, 1991.

[51] H. Wenzel, R. Güther, A. M. Shams-Zadeh-Amiri, and P. Bienstman, "A comparative study of higher order Bragg gratings: Coupled-mode theory versus mode expansion modeling," *IEEE J. Quantum Electron.*, vol. 42, no. 1, pp. 64–70, 2006.

[52] H.-J. Wünsche, M. Radziunas, S. Bauer, O. Brox, and B. Sartorius, "Modeling of mode control and noise in self-pulsating phaseCOMB lasers," *IEEE J. Sel. Top. Quantum Electron.*, vol. 9, no. 3, pp. 857–864, 2003.

[53] L. Landau and E. M. Lifshitz, *Electrodynamics of Continuous Media (Volume 8 of A Course of Theoretical Physics).* Addison-Wesley, 1982.

[54] H. Wenzel, "Basic aspects of high-power semiconductor laser simulation," *IEEE J. Sel. Top. Quantum Electron.*, vol. 19, no. 5, p. 1502913, 2013.

[55] H. Wenzel, F. Bugge, M. Dallmer, F. Dittmar, J. Fricke, K. H. Hasler, and G. Erbert, "Fundamental-lateral mode stabilized high-power ridge-waveguide lasers with a low beam divergence," *IEEE Photonics Technol. Lett.*, vol. 20, no. 3, pp. 214–216, 2008.

[56] H. Wenzel, G. Erbert, and P. M. Enders, "Improved theory of the refractive-index change in quantum-well lasers," *IEEE J. Sel. Top. Quantum Electron.*, vol. 5, no. 3, pp. 637–642, 1999.

[57] W. B. Joyce, "Current-crowded carrier confinement in double-heterostructure lasers," *J. Appl. Phys.*, vol. 51, no. 53, pp. 2394–7235, 1980.

[58] "WIAS-TeSCA: Two-dimensional semiconductor analysis package." *http://wias-berlin.de/software/tesca/*, accessed: 2020-01-01.

[59] K. Seeger, *Semiconductor Physics.* Springer Science & Business Media, 2013.

[60] J. Piprek, *Semiconductor Optoelectronic Devices.* San Diego: Academic Press, Elsevier, 1rst ed., 2003.

[61] A. H. Marshak and C. M. Van Vliet, "Electrical current and carrier density in degenerate materials with nonuniform band structure," *Proc. IEEE*, vol. 72, no. 2, pp. 148–164, 1984.

[62] W. B. Joyce, "Carrier transport in double-heterostructure active layers," *J. Appl. Phys.*, vol. 53, no. 11, pp. 7235–7239, 1982.

[63] W. B. Joyce and R. W. Dixon, "Analytic approximations for the Fermi energy of an ideal Fermi gas," *Appl. Phys. Lett.*, vol. 31, no. 5, pp. 354–356, 1977.

[64] W. B. Joyce, "Role of the conductivity of the confining layers in DH-laser spatial hole burning effects," *IEEE J. Quantum Electron.*, vol. 18, no. 12, pp. 2005–2009, 1982.

[65] T. Grasser, T.-W. Tang, H. Kosina, and S. Selberherr, "A review of hydrodynamic and energy-transport models for semiconductor device simulation," *Proc. IEEE*, vol. 91, no. 2, pp. 251–274, 2003.

[66] Y. Liu, W.-C. Ng, K. D. Choquette, and K. Hess, "Numerical investigation of self-heating effects of surface-emitting lasers," *IEEE J. Quantum Electron.*, vol. 41, no. 1, pp. 15–25, 2005.

[67] G. Albinus, H. Gajewski, and R. Hünlich, "Thermodynamic design of energy models of semiconductor devices," *Nonlinearity*, vol. 15, no. 2, p. 367, 2002.

[68] B. S. Shastry, "Thermopower in Correlated Systems," in *New Mater. Thermoelectr. Appl. Theory Exp.*, pp. 25–29, Dordrecht: Springer Netherlands, 2013.

[69] J. Cai and G. D. Mahan, "Effective Seebeck coefficient for semiconductors," *Phys. Rev. B*, vol. 74, p. 075201, 2006.

[70] J. Pomplun, H. Wenzel, S. Burger, L. Zschiedrich, M. R. Rozova, F. Schmidt, P. Crump, H. Ekhteraei, C. M. Schultz, and E. Götz, "Thermo-optical simulation of high-power diode lasers," *Proc. SPIE*, vol. 8255, p. 825510, 2012.

[71] J. Rieprich, M. Winterfeldt, R. Kernke, J. W. Tomm, and P. Crump, "Chip-carrier thermal barrier and its impact on lateral thermal lens profile and beam parameter product in high power broad area lasers," *J. Appl. Phys.*, vol. 123, p. 125703, 2018.

[72] L. C. Evans, *Partial Differential Equations. Graduate Studies in Mathematics.* Berkeley: American Mathematical Society, 19 ed., 1997.

[73] A. Klehr, H. Wenzel, O. Brox, S. Schwertfeger, R. Staske, and G. Erbert, "Dynamics of a gain-switched distributed feedback ridge waveguide laser in nanoseconds time scale under very high current injection conditions," *Opt. Express*, vol. 21, no. 3, pp. 2777–2786, 2013.

[74] A. Klehr, A. Liero, H. Wenzel, A. Zeghuzi, J. Fricke, R. Staske, and A. Knigge, "Pico- and nanosecond investigations of the lateral nearfield of broad area lasers under pulsed high-current excitation," *Proc. SPIE*, vol. 10553, p. 105530K, 2018.

[75] R. MacKenzie, J. J. Lim, S. Bull, S. Sujecki, A. J. Kent, and E. C. Larkins, "The impact of hot-phonons on the performance of 1.3 μm dilute nitride edge-emitting quantum well lasers," *J. Phys. Conf. Ser.*, vol. 92, p. 012068, 2007.

[76] C.-Y. Tsai, C.-Y. Tsai, Y.-H. Lo, and L. F. Eastman, "Carrier energy relaxation time in quantum-well lasers," *IEEE J. Quantum Electron.*, vol. 31, no. 12, pp. 2148–2158, 1995.

[77] E. T. Swartz and R. O. Pohl, "Thermal boundary resistance," *Rev. Mod. Phys.*, vol. 61, no. 3, pp. 605–668, 1989.

[78] R. MacKenzie, J. J. Lim, S. Bull, S. Sujecki, and E. C. Larkins, "Inclusion of thermal boundary resistance in the simulation of high-power 980 nm ridge waveguide lasers," *Opt. Quantum Electron.*, vol. 40, pp. 373–377, 2008.

[79] A. Knigge, A. Klehr, H. Wenzel, A. Zeghuzi, J. Fricke, A. Maaßdorf, A. Liero, and G. Tränkle, "Wavelength-stabilized high-pulse-power laser diodes for automotive LiDAR," *Phys. Status Solidi A*, vol. 215, no. 8, p. 1700439, 2018.

[80] J. Piprek and Z.-M. Li, "What causes the pulse power saturation of GaAs-based broad-area lasers," *IEEE Photonics Technol. Lett.*, vol. 30, no. 10, pp. 963–966, 2018.

[81] L. Borruel, J. Arias, B. Romero, and I. Esquivias, "Incorporation of carrier capture and escape processes into a self-consistent cw model for Quantum Well lasers," *Microelectronics J.*, vol. 34, pp. 675–677, 2003.

[82] D. Mehuys, R. Lang, M. Mittelstein, J. Salzman, and A. Yariv, "Self-stabilized nonlinear lateral modes of broad area lasers," *IEEE J. Quantum Electron.*, vol. 23, no. 11, pp. 1909–1920, 1987.

[83] R. J. Lang, A. G. Larsson, and G. Jeffrey, "Lateral modes of broad area semiconductor lasers: Theory and experiment," *IEEE J. Quantum Electron.*, vol. 27, no. 3, pp. 312–320, 1991.

[84] J. V. Moloney, "Semiconductor laser device modeling," *AIP Conf. Proc.*, vol. 548, pp. 149–172, 2000.

[85] E. Gehrig and O. Hess, *Spatio-temporal dynamics and quantum fluctuations in semiconductor lasers*. Berlin, Heidelberg, New York: Springer Science & Business Media, 1rst ed., 2003.

[86] K. Böhringer, *Microscopic spatio-temporal dynamics of semiconductor quantum well lasers and amplifiers*. PhD thesis, Deutsches Zentrum für Luft- und Raumfahrt, 2007.

[87] O. Hess, "Spatio-temporal complexity in multi-stripe and broad-area semiconductor lasers," *Chaos, Solitons and Fractals*, vol. 4, no. 8/9, pp. 1597–1618, 1994.

[88] H. Wenzel, P. Crump, H. Ekhteraei, C. Schultz, J. Pomplun, S. Burger, L. Zschiedrich, F. Schmidt, and G. Erbert, "Theoretical and experimental analysis of the lateral modes of high-power broad-area lasers," *Proc. NUSOD*, vol. Sep. 2011, no. 1, pp. 143–144, 2011.

[89] N. Stelmakh, S. Member, and M. Flowers, "Measurement of spatial modes of broad-area diode lasers with 1-GHz resolution grating spectrometer," *IEEE Photonics Technol. Lett.*, vol. 18, no. 15, pp. 1618–1620, 2006.

[90] P. Crump, M. Ekterai, C. M. Schultz, G. Erbert, and G. Tränkle, "Studies of limitations to lateral brightness in high power diode lasers using spectrally-resolved mode profiles," *IEEE Int. Semicond. Laser Conf.*, pp. 23–24, 2014.

[91] G. P. Agrawal, *Nonliear Fiber Optics*. San Diego: Academic Press, 3rd ed., 2004.

[92] P. Crump, M. Winterfeldt, J. Decker, M. Ekterai, J. Fricke, S. Knigge, A. Maaßdorf, and G. Erbert, "Novel approaches to increasing the brightness of broad area lasers," *Proc. SPIE*, vol. 9767, p. 97671L, 2016.

[93] C. Holly, X. Liu, S. Heinemann, S. McDougall, and H. Zimer, "Influence of lateral refractive index profiles on the divergence angle of gain-guided broad-area laser diode bars," *IEEE Photonics Conf.*, 2018.

[94] S. Blaaberg, P. M. Petersen, and B. Tromborg, "Structure, stability, and spectra of lateral modes of a broad-area semiconductor laser," *IEEE J. Quantum Electron.*, vol. 43, no. 11, pp. 959–973, 2007.

[95] H. S. J. Sommers, "Experimental properties of injection lasers: modal distribution of laser power," *J. Appl. Phys.*, vol. 44, p. 1263, 1973.

[96] F. C. Prince, T. J. S. Mattos, N. B. Patel, D. Kasemset, and C.-S. Hong, "Waveguiding, spectral, and threshold properties of a stripe geometry single quantum well laser," *IEEE J. Quantum Electron.*, vol. QE-21, no. 6, pp. 634–639, 1985.

[97] A. J. Benett, R. D. Clayton, and J. M. Xu, "Above-threshold longitudinal profiling of carrier nonpinning and spatial modulation in asymmetric cavity lasers," *J. Appl. Phys.*, vol. 83, no. 7, p. 3784, 1998.

[98] H. An, Y. Xiong, C.-L. J. Jiang, B. Schmidt, and G. Treusch, "Methods for slow axis beam quality improvement of high power broad area diode lasers," *Proc. SPIE*, vol. 8965, p. 89650U, 2014.

[99] Y. Yamagata, Y. Yamada, M. Muto, S. Sato, R. Nogawa, A. Sakamoto, and M. Yamaguchi, "915nm high-power broad area laser diodes with ultra-small optical confinement based on asymmetric decoupled confinement heterostructure (ADCH)," *Proc. SPIE*, vol. 9348, p. 93480F, 2015.

[100] J. G. Bai, P. Leisher, S. Zhang, S. Elim, M. Grimshaw, C. Bai, L. Bintz, D. Dawson, L. Bao, J. Wang, M. DeVito, R. Martinsen, and J. Haden, "Mitigation of thermal lensing effect as a brightness limitation of high-power broad area diode lasers," *Proc. SPIE*, vol. 7953, pp. 79531F–1–7, 2011.

[101] W. Sun, R. Pathak, G. Campbell, H. Eppich, J. H. Jacob, A. Chin, and J. Fryer, "Higher brightness laser diodes with smaller slow axis divergence," *Proc. SPIE*, vol. 8605, p. 86050D, 2013.

[102] J. Piprek, "Inverse thermal lens effects on the far-field blooming of broad area laser diodes," *IEEE Photonics Technol. Lett.*, vol. 25, no. 10, pp. 958–960, 2013.

[103] H.-C. Eckstein, U. Zeitner, A. Tünnermann, C. Lauer, and U. Strauß, "Numerical simulation and optimization of microstructured high brightness broad area laser diodes," *Proc. SPIE*, vol. 9382, p. 93821H, 2015.

[104] D. Botez and D. R. Scifres, eds., *Diode Laser Arrays.* Cambridge University Press, 1994.

[105] M. Radziunas, M. Botey, R. Herrero, and K. Staliunas, "Intrinsic beam shaping mechanism in spatially modulated broad area semiconductor amplifiers," *Appl. Phys. Lett.*, vol. 103, no. 13, p. 132101, 2013.

[106] C. Lauer, H. König, U. Strauß, and A. Bachmann, "Semiconductor Laser Diode US 9, 722, 394 B2."

[107] Z. Chen, L. Bao, J. Bai, M. Grimshaw, R. Martinsen, M. Devito, and J. Haden, "Performance limitation and mitigation of longitudinal spatial hole burning in high-power diode lasers," *Proc. SPIE*, vol. 8277, p. 82771J, 2012.

[108] C. Holly, *Modeling of the lateral emission characteristics of high-power edge-emitting semiconductor lasers.* PhD thesis, Technische Hochschule Aachen, 2019.

[109] K. Staliunas, R. Herrero, and R. Vilaseca, "Subdiffraction and spatial filtering due to periodic spatial modulation of the gain-loss profile," *Phys. Rev. A*, vol. 80, no. 1, p. 013821, 2009.

[110] D. Botez and L. J. Mawst, "Phase-locked laser arrays revisited," *IEEE Circuits Devices Mag.*, vol. 12, no. 6, pp. 25–32, 1996.

[111] D. Botez and D. E. Ackley, "Phase-locked arrays of semiconductor diode lasers," *IEEE Circuits Devices Mag.*, vol. 2, no. 1, pp. 8–17, 1986.

[112] N. W. Carlsson, *Monolithic Diode-Layer Arrays (Springer Series in Electronics and Photonics 33).* Springer-Verlag Berlin Heidelberg, 1994.

[113] D. F. Siriani and K. D. Choquette, "Reduced loss and improved mode discrimination in resonant optical waveguide arrays," *Electron. Lett.*, vol. 48, no. 10, pp. 3–4, 2012.

[114] J. S. Major, D. Mehuys, and D. F. Welch, "11.5 W pulsed operation of antiguided laser diode array," *J. Chem. Inf. Model.*, vol. 53, no. 9, pp. 1689–1699, 2019.

[115] K. L. Chen and S. Wang, "Analysis of symmetric y-junction laser arrays with uniform near-field distribution," *Electron. Lett.*, vol. 22, no. 12, pp. 644–645, 1986.

[116] P. D. V. Eijk, M. Reglat, G. Vassilieff, G. J. M. Krijnen, A. Driessen, and A. J. Mouthaan, "Analysis of the modal behavior of an antiguide diode laser array with Talbot filter," *J. Light. Technol.*, vol. 9, no. 5, pp. 629–634, 1991.

[117] H. Yang, L. J. Mawst, M. Nesnidal, J. Lopez, A. Bhattacharya, and D. Botez, "10 W near-diffraction-limited peak pulsed power from Al-free phase-locked antiguided arrays," *Electron. Lett.*, vol. 33, no. 2, pp. 136–137, 1997.

[118] J. P. Koester, M. Radziunas, A. Zeghuzi, H. Wenzel, and A. Knigge, "Simulation and design of a compact GaAs based tunable dual-wavelength diode laser system," *Opt. Quantum Electron.*, vol. 51, no. 10, pp. 1–12, 2019.

[119] T. Hao, J. Song, R. Liptak, and P. O. Leisher, "Experimental verification of longitudinal spatial hole burning in high-power diode lasers," *Proc. SPIE*, vol. 9081, p. 90810U, 2014.

[120] M. Peters, V. Rossin, M. Everett, and E. Zucker, "High power, high efficiency laser diodes at JDSU," *Proc. SPIE*, vol. 6456, p. 64560G, 2007.

[121] U. Menzel, "Self-consistent calculation of facet heating in asymmetrically coated edge emitting diode lasers," *Semicond. Sci. Technol.*, vol. 13, pp. 265–276, 1998.

[122] S. Gehrsitz, F. K. Reinhart, C. Gourgon, N. Herres, A. Vonlanthen, and H. Sigg, "The refractive index of AlxGa1-xAs below the band gap: Accurate determination and empirical modeling," *J. Appl. Phys.*, vol. 87, no. 11, pp. 7825–7837, 2000.

[123] L. A. Coldren, S. Corzine, and M. L. Masanovic, *Diode Lasers and Photonic Integrated Circuits.* John Wiley & Sons, 2nd ed., 2012.

[124] K. Unger, "Spontane und induzierte Emission in Laserdioden," *Zeitschrift für Phys.*, vol. 207, pp. 322–331, 1967.

[125] J. Fuhrmann and T. Steckenbach, "pdelib: A finite volume and finite element toolbox for partial differential equations.." *http://www.wias-berlin.de/software/pdelib/*, accessed: 2020-01-01.

[126] R. Čiegis and M. Radziunas, "Effective numerical integration of traveling wave model for edge-emitting broad-area semiconductor lasers and amplifiers," *Math. Model. Anal.*, vol. 15, no. 4, pp. 409–430, 2010.

[127] R. Čiegis, M. Radziunas, and M. Lichtner, "Numerical algorithms for simulation of multisection lasers by using traveling wave model," *Math. Model. Anal.*, vol. 13, no. 3, pp. 327–348, 2008.

Selbstständigkeitserklärung

Ich erkläre, dass ich die Dissertation selbstständig und nur unter Verwendung der von mir gemäß § 7 Abs. 3 der Promotionsordnung der Mathematisch-Naturwissenschaftlichen Fakultät, veröffentlicht im amtlichen Mitteilungsblatt der Humboldt-Universität zu Berlin Nr. 42/2018 am 11.07.2018 angegebenen Hilfsmittel angefertigt habe.

Berlin, den 13.02.2020

Anissa Zeghuzi

Innovationen mit Mikrowellen und Licht
Forschungsberichte aus dem Ferdinand-Braun-Institut, Leibniz-Institut für Höchstfrequenztechnik

Herausgeber: Prof. Dr. G. Tränkle, Prof. Dr.-Ing. W. Heinrich

Band 1: **Thorsten Tischler**
Die Perfectly-Matched-Layer-Randbedingung in der Finite-Differenzen-Methode im Frequenzbereich: Implementierung und Einsatzbereiche
ISBN: 3-86537-113-2, 19,00 EUR, 144 Seiten

Band 2: **Friedrich Lenk**
Monolithische GaAs FET- und HBT-Oszillatoren mit verbesserter Transistormodellierung
ISBN: 3-86537-107-8, 19,00 EUR, 140 Seiten

Band 3: **R. Doerner, M. Rudolph (eds.)**
Selected Topics on Microwave Measurements, Noise in Devices and Circuits, and Transistor Modeling
ISBN: 3-86537-328-3, 19,00 EUR, 130 Seiten

Band 4: **Matthias Schott**
Methoden zur Phasenrauschverbesserung von monolithischen Millimeterwellen-Oszillatoren
ISBN: 978-3-86727-774-0, 19,00 EUR, 134 Seiten

Band 5: **Katrin Paschke**
Hochleistungsdiodenlaser hoher spektraler Strahldichte mit geneigtem Bragg-Gitter als Modenfilter (α-DFB-Laser)
ISBN: 978-3-86727-775-7, 19,00 EUR, 128 Seiten

Band 6: **Andre Maaßdorf**
Entwicklung von GaAs-basierten Heterostruktur-Bipolartransistoren (HBTs) für Mikrowellenleistungszellen
ISBN: 978-3-86727-743-3, 23,00 EUR, 154 Seiten

Band 7: **Prodyut Kumar Talukder**
Finite-Difference-Frequency-Domain Simulation of Electrically Large Microwave Structures using PML and Internal Ports
ISBN: 978-3-86955-067-1, 19,00 EUR, 138 Seiten

Band 8: **Ibrahim Khalil**
Intermodulation Distortion in GaN HEMT
ISBN: 978-3-86955-188-3, 23,00 EUR, 158 Seiten

Band 9: **Martin Maiwald**
Halbleiterlaser basierte Mikrosystemlichtquellen für die Raman-Spektroskopie
ISBN: 978-3-86955-184-5, 19,00 EUR, 134 Seiten

Band 10: **Jens Flucke**
Mikrowellen-Schaltverstärker in GaN- und GaAs-Technologie Designgrundlagen und Komponenten
ISBN: 978-3-86955-304-7, 21,00 EUR, 122 Seiten

Cuvillier Verlag
Internationaler wissenschaftlicher Fachverlag

Innovationen mit Mikrowellen und Licht
Forschungsberichte aus dem Ferdinand-Braun-Institut, Leibniz-Institut für Höchstfrequenztechnik

Herausgeber: Prof. Dr. G. Tränkle, Prof. Dr.-Ing. W. Heinrich

Band 11: **Harald Klockenhoff**
Optimiertes Design von Mikrowellen-Leistungstransistoren und Verstärkern im X-Band
ISBN: 978-3-86955-391-7, 26,75 EUR, 130 Seiten

Band 12: **Reza Pazirandeh**
Monolithische GaAs FET- und HBT-Oszillatoren mit verbesserter Transistormodellierung
ISBN: 978-3-86955-107-8, 19,00 EUR, 140 Seiten

Band 13: **Tomas Krämer**
High-Speed InP Heterojunction Bipolar Transistors and Integrated Circuits in Transferred Substrate Technology
ISBN: 978-3-86955-393-1, 21,70 EUR, 140 Seiten

Band 14: **Phuong Thanh Nguyen**
Investigation of spectral characteristics of solitary diode lasers with integrated grating resonator
ISBN: 978-3-86955-651-2, 24,00 EUR, 156 Seiten

Band 15: **Sina Riecke**
Flexible Generation of Picosecond Laser Pulses in the Infrared and Green Spectral Range by Gain-Switching of Semiconductor Lasers
ISBN: 978-3-86955-652-9, 22,60 EUR, 136 Seiten

Band 16: **Christian Hennig**
Hydrid-Gasphasenepitaxie von versetzungsarmen und freistehenden GaN-Schichten
ISBN: 978-3-86955-822-6, 27,00 EUR, 162 Seiten

Band 17: **Tim Wernicke**
Wachstum von nicht- und semipolaren InAlGaN-Heterostrukturen für hocheffiziente Licht-Emitter
ISBN: 978-3-86955-881-3, 23,40 EUR, 138 Seiten

Band 18: **Andreas Wentzel**
Klasse-S Mikrowellen-Leistungsverstärker mit GaN-Transistoren
ISBN: 978-3-86955-897-4, 29,65 EUR, 172 Seiten

Band 19: **Veit Hoffmann**
MOVPE growth and characterization of (In,Ga)N quantum structures for laser diodes emitting at 440 nm
ISBN: 978-3-86955-989-6, 18,00 EUR, 118 Seiten

Band 20: **Ahmad Ibrahim Bawamia**
Improvement of the beam quality of high-power broad area semiconductor diode lasers by means of an external resonator
ISBN: 978-3-95404-065-0, 21,00 EUR, 126 Seiten

Cuvillier Verlag
Internationaler wissenschaftlicher Fachverlag

Innovationen mit Mikrowellen und Licht
Forschungsberichte aus dem Ferdinand-Braun-Institut, Leibniz-Institut für Höchstfrequenztechnik

Herausgeber: Prof. Dr. G. Tränkle, Prof. Dr.-Ing. W. Heinrich

Band 21: **Agnietzka Pietrzak**
Realization of High Power Diode Lasers with Extremely Narrow Vertical Divergence
ISBN: 978-3-95404-066-7, 27,40 EUR, 144 Seiten

Band 22: **Eldad Bahat-Treidel**
GaN-based HEMTs for High Voltage Operation
Design, Technology and Characterization
ISBN: 978-3-95404-094-0, 41,10 EUR, 220 Seiten

Band 23: **Ponky Ivo**
AlGaN/GaN HEMTs Reliability:
Degradation Modes and Anslysis
ISBN: 978-3-95404-259-3, 23,55 EUR, 132 Seiten

Band 24: **Stefan Spießberger**
Compact Semiconductor-Based Laser Sources
with Narrow Linewidth and High Output Power
ISBN: 978-3-95404-261-6, 24,15 EUR, 140 Seiten

Band 25: **Silvio Kühn**
Mikrowellenoszillatoren für die Erzeugung von atmosphärischen Mikroplasmen
ISBN: 978-3-95404-378-1, 21,85 EUR, 112 Seiten

Band 26: **Sven Schwertfeger**
Experimentelle Untersuchung der Modensynchronisation in Multisegment-Laserdioden zur Erzeugung kurzer optischer Pulse bei einer Wellenlänge von 920 nm
ISBN: 978-3-95404-471-9, 29,45 EUR, 150 Seiten

Band 27: **Christoph Matthias Schultz**
Analysis and mitigation of the factors limiting the effiency of high power distributed feedback diode lasers
ISBN: 978-3-95404-521-1, 68,40 EUR, 388 Seiten

Band 28: **Luca Redaelli**
Design and fabrication of GaN-based laser diodes for single-mode and narrow-linewidth applications
ISBN: 978-3-95404-586-0, 29,70 EUR, 176 Seiten

Band 29: **Martin Spreemann**
Resonatorkonzepte für Hochleistungs-Diodenlaser
mit ausgedehnten lateralen Dimensionen
ISBN: 978-3-95404-628-7, 25,15 EUR, 128 Seiten

Cuvillier Verlag
Internationaler wissenschaftlicher Fachverlag

Innovationen mit Mikrowellen und Licht
Forschungsberichte aus dem Ferdinand-Braun-Institut, Leibniz-Institut für Höchstfrequenztechnik

Herausgeber: Prof. Dr. G. Tränkle, Prof. Dr.-Ing. W. Heinrich

Band 30: **Christian Fiebig**
Diodenlaser mit Trapezstruktur und hoher Brillanz für die Realisierung einer Frequenzkonversion auf einer mikro-optischen Bank
ISBN: 978-3-95404-690-4, 26,30 EUR, 140 Seiten

Band 31: **Viola Küller**
Versetzungsreduzierte AlN- und AlGaN-Schichten als Basis für UV LEDs
ISBN: 978-3-95404-741-3, 34,40 EUR, 164 Seiten

Band 32: **Daniel Jedrzejczyk**
Efficient frequency doubling of near-infrared diode lasers using quasi phase-matched waveguides
ISBN: 978-3-95404-958-5, 27,90 EUR, 134 Seiten

Band 33: **Sylvia Hagedorn**
Hybrid-Gasphasenepitaxie zur Herstellung von Aluminiumgalliumnitrid
ISBN: 978-3-95404-985-1, 38,00 EUR, 176 Seiten

Band 34: **Alexander Kravets**
Advanced Silicon MMICs for mm-Wave Automotive Radar Front-Ends
ISBN: 978-3-95404-986-8, 31,90 EUR, 156 Seiten

Band 35: **David Feise**
Longitudinale Modenfilter für Kantanemitter im roten Spektralbereich
ISBN: 978-3-7369-9116-3, 39,20 EUR, 168 Seiten

Band 36: **Ksenia Nosaeva**
Indium phosphide HBT in thermally optimized periphery for applications up to 300GHZ
ISBN: 978-3-7369-287-0, 42,00 EUR, 154 Seiten

Band 37: **Muhammad Maruf Hossain**
Signal Generation for Millimeter Wave and THZ Applications in InP-DHBT and InP-on-BiCMOS Technologies
ISBN: 978-3-7369-9335-8, 35,60 EUR, 136 Seiten

Band 38: **Sirinpa Monayakul**
Development of Sub-mm Wave Flip-Chip Interconnect
ISBN: 978-3-7369-9410-2, 44,00 EUR, 146 Seiten

Band 39: **Moritz Brendel**
Charakterisierung und Optimierung von (Al, Ga) N-basierten UV-Photodetektoren
ISBN: 978-3-7369-9465-2, 49,90 EUR, 196 Seiten

Band 40: **Erdenetsetseg Luvsandamdin**
Development of micro-integrated diode lasers for precision quantum optics experiments in space
ISBN: 978-3-7369-9479-9, 39,00 EUR, 126 Seiten

Internationaler wissenschaftlicher Fachverlag

Innovationen mit Mikrowellen und Licht
Forschungsberichte aus dem Ferdinand-Braun-Institut, Leibniz-Institut für Höchstfrequenztechnik

Herausgeber: Prof. Dr. G. Tränkle, Prof. Dr.-Ing. W. Heinrich

Band 41: **Thi Nghiem Vu**
Development and analysis of diode laser ns-MOPA systems for high peak power application
ISBN: 978-3-7369-9480-5, 38,80 EUR, 138 Seiten

Band 42: **Christian Bansleben**
Differentieller Mikrowellen-Leistungsoszillator für die Realisierung ultrakompakter Plasmaquellen in Matrixanordnung
ISBN: 978-3-7369-9530-7, 34,90 EUR, 136 Seiten

Band 43: **Martin Winterfeldt**
Investigation of slow-axis beam quality degradation in high-power broad area diode lasers
ISBN: 978-3-7369-9733-2, 39,90 EUR, 158 Seiten

Band 44: **Jonathan Decker**
Investigation of monolithically integrated spectral stabilization in high-brightness broad area diode lasers
ISBN: 978-3-7369-9798-1, 49,50 EUR, 174 Seiten

Band 45: **Andreea Cristina Andrei**
Untersuchung und Optimierung robuster und hochlinearer rauscharmer Verstärker in GaN-Technologie
ISBN: 978-3-7369-9810-0, 39,90 EUR, 148 Seiten

Band 46: **Peng Luo**
GaN HEMT Modeling Including Trapping Effects Based on Chalmers Model and Pulsed S-Parameter Measurements
ISBN: 978-3-7369-9906-0, 48,00 EUR, 160 Seiten

Band 47: **Jörg Jeschke**
Entwicklung von optisch pumpbaren UVC-Lasern auf AlGaN-Basis Chalmers Model and Pulsed S-Parameter Measurements
ISBN: 978-3-7369-9918-3, 44,90 EUR, 176 Seiten

Band 48: **Nikolai Wolff**
Wideband GaN Microwave Power Amplifiers with Class-G Supply Modulation
ISBN: 978-3-7369-9931-2, 44,90 EUR, 170 Seiten

Band 49: **Carlo Frevert**
Optimization of broad-area GaAs diode lasers for high powers and high efficiences in the temperature range 200-220 K
ISBN: 978-3-7369-9944-2, 44,90 EUR, 174 Seiten

Band 50: **Bassem Arar**
GaAs-based components for photonic integrated circuits
ISBN: 978-3-7369-9976-3, 43,60 EUR, 152 Seiten

Internationaler wissenschaftlicher Fachverlag

Innovationen mit Mikrowellen und Licht
Forschungsberichte aus dem Ferdinand-Braun-Institut, Leibniz-Institut für Höchstfrequenztechnik

Herausgeber: Prof. Dr. G. Tränkle, Prof. Dr.-Ing. W. Heinrich

Band 51: **Mahmoud Tawfieq**
Development and characterisation of a diode laser based tunable high-power MOPA system
ISBN: 978-3-7369-9983-1, 44,90 EUR, 170 Seiten

Band 52: **Sebastian Preis**
Hocheffiziente frequenzagile Mikrowellen-Leistungsverstärker auf Basis von Verbindungshalbleitern und Ferroelektrika
ISBN: 978-3-7369-7004-5, 54,00 EUR, 136 Seiten

Band 53: **Simon Fleischmann**
Materialaspekte der Hydridgasphasenepitaxie von Aluminiumgalliumnitrid
ISBN: 978-3-7369-7029-8, 41,80 EUR, 160 Seiten

Band 54: **Erhan Ersoy**
Optimierung von koplanaren GaN-MMIC-Leistungsverstärkern im X-Band
ISBN: 978-3-7369-7157-8, 39,90 EUR, 154 Seiten

Band 55: **Simon Rauch**
Lateral emission characteristics of high-power broad-area lasers subject to external optical feedback
ISBN: 978-3-7369-7168-5, 29,90 EUR, 112 Seiten

Band 56: **Frank Dittmar**
Untersuchung der Strahlgüte von brillanten Hochleistungs-Trapezlasern für den Wellenlängenbereich bei 808 nm
ISBN: 978-3-7369-7167-7, 44,90 EUR, 176 Seiten

Band 57: **Florian Hühn**
Flexibler Modulator und Digitalverstärker-MMIC für den energieeffizienten Betrieb einer digitalen Sendekette im GHz-Bereich
ISBN: 978-3-7369-7190-5, 29,90 EUR, 162 Seiten

Band 58: **Dimitri Stoppel**
Interconnection development for InP-HBT terahertz circuits
ISBN: 978-3-7369-7204-9, 44,90 EUR, 156 Seiten

Cuvillier Verlag
Internationaler wissenschaftlicher Fachverlag

www.ingramcontent.com/pod-product-compliance
Ingram Content Group UK Ltd.
Pitfield, Milton Keynes, MK11 3LW, UK
UKHW022001190726
13853UKWH00004B/1655